Microtectonics along the Western Edge of the Blue Ridge, Maryland and Virginia

The Johns Hopkins University
STUDIES IN GEOLOGY

*No. 20 Microtectonics along the Western Edge
of the Blue Ridge, Maryland and Virginia*

Microtectonics along the Western Edge of the Blue Ridge, Maryland and Virginia

ERNST CLOOS

The Johns Hopkins Press

BALTIMORE AND LONDON

The Johns Hopkins Press, Baltimore, Maryland 21218
The Johns Hopkins Press Ltd., London

Library of Congress Catalog Card Number 77-156828
ISBN 0-8018-1281-X

To the memory of
ROBERT BALK

Contents

Tables

List of Figures

List of Plates

LIST OF PLATES (*continued*)

Acknowledgments

The author most gratefully acknowledges grants by the Geological Society of America for the continuation of the oölite study south of the Potomac River and by the National Science Foundation to enlarge the scope of the study to include the determination of the deformation plane over a larger area and to complete the study.

Many individuals have lent a helping hand: Mrs. A. Z. Huebner and Mrs. John S. Wickham were most faithful and efficient assistants. Plates 1 and 2 and the diagrams were drafted by Karen Wickham. John Wickham covered a great deal of terrane in the Front Royal area, mapped the Front Royal thrusts and contributed much information. Mrs. B. Schrago typed the manuscript and assisted in many ways. Thin sections and material preparations were by Michael Krasnokutsky. The Administration of Shenandoah National Park furnished most valuable and vital assistance. The manuscript was greatly improved by innumerable suggestions by Agnes Creagh and David Elliott. Charles H. Weber prepared the plates for printing.

Microtectonics along the
Western Edge of the Blue Ridge,
Maryland and Virginia

Abstract

The South Mountain plan (Cloos 1947) was traced southward across the Potomac River to Lexington, Virginia, without much modification. Thrusts at Front Royal and several faults complicate the picture locally.

In addition to the determination of strain in three oölite belts measurements were made on orientations of lineations, striations on slickensided surfaces, and mineral growth in fractures. The long ellipsoid axes, lineation, striae, and growths determine a deformation plane generally normal to fold axes which is incredibly uniform from Pennsylvania to Lexington and in the stratigraphic column, including the basement gneiss and the Silurian of Massanutten Mountain. In that column and from west to east deformation intensity increases gradually across the Great Valley and the Blue Ridge.

In a zone about 30 miles wide, basement and paleozoic sediments are welded together by the structures and orientations that cut across basement gneiss, Catoctin Volcanics, Chilhowee Formations and carbonates up to the Silurian or higher. Small breaks occur and the upturned western edge of the crystalline basement is folded together with its cover. This does not exclude gravity tectonics higher in the section or west of the Great Valley.

The consistent pattern calls for mechanisms and causes that affect the area from the end of South Mountain in Pennsylvania to Lexington, Virginia, rather uniformly and cannot be explained by local events. These are readily recognized where they occur.

The crystallines have moved upward and northwest, affecting the sedimentary cover in a sequence of events that began with an activation of the basement at depth and then a rise and westward creep with a push from "behind" and "within" the crystalline axis. This could be due to convection, ocean floor spreading, or any other mechanism that produced the multitude of structures and the uniform deformation pattern.

I

Introduction

The discovery of deformed oölites in Maryland led to a systematic study that I reported some years ago (E. Cloos 1947). The lower Paleozoic limestones were shown to be folded and deformed together with the underlying upper Precambrian Volcanics and igneous complex. An axial-plane flow cleavage of the South Mountain fold cuts through Catoctin Volcanics, the clastic sediments of the Chilhowee Group and upward into Cambrian and Ordovician carbonates. Cleavage, however, is difficult to see in some of the limestones and massive Precambrian greenstones and the Antietam Quartzite. Thus cleavage is not everywhere visible nor is it always in the same attitude, because it depends on lithologies and individual folds. Cleavage is common, however, in all suitable rocks; even the Martinsburg (Ordovician) and Jennings (Devonian) show a very distinct and dominant cleavage.

In the cleavage surfaces there is a consistent lineation normal to fold axes. This lineation is very prominent in the South Mountain fold (Cloos 1946, 1947, 1951; Nickelsen 1956; Reed 1955) and extends upward into the limestones; it parallels the long axes of the deformed oöids in oölites. In some zones it also appears in the Beekmantown, but not uniformly. West of South Mountain in the Great Valley that lineation disappears at the tectonite front (Fellows 1943) and does not occur in the Martinsburg or Jennings Formations.

Since 1958 the oölite study has been continued southward to the James River, Virginia, and many new oölite localities have been established. The earlier study was fairly easy because the geology of Washington County, Maryland, had just been mapped, but to the south there was not as much information available, and progress was slow. In 1966 we discovered that in addition to the relationship between lineation, oölite extension, and cleavage, the striation on slickensided surfaces, and growths of fibrous minerals in fractures were also related. If the ac plane is normal to the fold axes that same surface contains the lineation a, the long axes of the deformed oöids and the maxima for striations and mineral growths. This suggests a similar and uniform deformation plan for all elements, stratigraphically from the Precambrian at least to the Silurian.

This finding prompted a systematic search for the orientation of this de-

1

formation plane—*ac*—in a larger area to the south including the Blue Ridge of Virginia almost to the James River and with as much of its foreland to the west as was necessary to establish the relationships.

An analysis of Appalachian folding must include the sedimentary column and the underlying basement on which it was deposited. Across the Appalachian belt the stratigraphic column extends from the Precambrian basement complex to the Pennsylvanian—a vertical distance of at least 25,000 to 30,000 feet, or about 6 miles. In time this column spans 400 million years. Most critical for tectonic analysis is the upper portion of the basement and the lower half of the sedimentary column.

The South Mountain–Blue Ridge uplift of the central Appalachians is one of the few areas where the relationship between crystalline (Precambrian) basement and overlying Paleozoic sedimentary rocks is preserved. Many contacts are still normal. In most folded belts the basement surface has been severed from the cover by faulting and thrusting, making it impossible to study the original relationships within the contact zone. In alpine sections, crystallines are faulted against sediments. In the Caledonides great thrusts separate the Moines and Lewesian from Torridonian or Cambrian, and in Sweden and Norway crystallines have overridden Silurian limestones. Along the eastern front of the Rocky Mountains great thrusts are well known. Even in the Jura the basement surface is a dislocation on which a more mobile cover has slipped off a more stable basement. In the Southern and Northern Applachians extensive thrust faults are well known (Reed 1970, p. 195).

This study aims to determine the relationship of the basement structures to those in the overlying sedimentary rocks and their regional changes. The regional approach is the necessary link between detailed measurements and observations and tectonic interpretations. I am trying to relate small-scale measurements to larger areas, because if single observations are not placed into a general context they remain collectors' items of limited geological meaning.

Several studies of quantitative strain determinations have been made on oölites, fossils, and pebbles. Strain analysis has been reviewed thoroughly and mathematically by Ramsay (1967).

Many published maps and sections show orientation and distribution of structure elements, folding styles, cleavages, lineations, or metamorphic grades, but deformation-intensity maps are scarce, because there are very few suitable areas where such studies can be undertaken, and still fewer rocks contain measurable strain markers such as oölites, pebbles, amygdules, or fossils over a large and tectonically critical area. Even so, very much more could be done than has been done.

The distance between the Potomac River and the James River is about 160 miles, and the width of the area is about 40 miles. Within this area it is difficult to cover a map evenly with meaningful information, and sampling became a major consideration. The area covers roughly 6,400 square miles. If the affected thickness of strata is 6 miles, at least 38,500 cubic miles of rock have been folded, moved, stressed, heated, recrystallized, compressed, thickened, and transported many miles to the northwest. In addition, the section consists of widely differing units ranging from crystalline gneisses and basic lava flows to limestones and shales. It is inconceivable that this immense block of earth crust was moved and altered homogeneously at the same rate or in the same manner.

A few simple facts emerge, however, and they are surprisingly consistent in the whole region, including South Mountain to the north. The directions of transport, the deformation plane (ac), and the change in deformation intensity are amazingly uniform. To determine these facts it was necessary to sample the information and treat it statistically in graphs, charts, diagrams, and maps as described below. Data from earlier studies are incorporated in Plate 1.

In spite of the time invested, the data collected, and the over-all effort, this study is a reconnaissance and probably points out more problems than it solves.

II

Definitions

Cleavage: "all types of secondary planar parallel fabric elements (other than coarse schistosity) which impart mechanical anisotropy to the rock without apparent loss of cohesion" (International Tectonic Dictionary, Dennis 1967).

Two cleavages are common: (1) flow cleavage, axial plane cleavage, regional cleavage (S_1); (2) fracture cleavage, a second cleavage that cuts (1), rarely accompanied by mineral rearrangement (S_2). Also called slip cleavage (Nickelsen 1956, pp. 246–57; Billings 1954, p. 339; Wickham 1969; Stevens 1959 [S_2]).

Bedding: S_0

Co-ordinates: a, b, c

ab plane is cleavage plane

b fold axis and S_0–S_1 intersection, mostly and statistically coincident over a wide area

c normal to ab

a in ab normal to b also a prominent lineation

Deformation plane: ab

Oöid ellipsoids: A—longest

B—intermediate $\}$ axes

C—shortest

Ratios: axial relations A/C, A/B, B/C.

Center cut: oöid cut through center.

Low or high cut: oöid cut below or above center.

Area 1: westernmost of three oölite belts on west side of Great Valley.

Area 2: central oölite belt along western foothills of Massanutten Mountain.

Area 3: oölite belt along foothills of Blue Ridge, between Massanutten Mountain and Blue Ridge.

Subareas 01 to 12: 18 segments of Skyline Drive and Blue Ridge Parkway used in statistical analysis of slickenside orientation (Fig. 24).

For oöid types see Table 1.

Epidosite: pod-shaped segregations of quartz and epidote. Common diameters measured in meters (Reed 1955, p. 879).

Diagrams: all were made with equal-area net presenting lower hemisphere.

5

Directions are shown as azimuths: North equal 0 or 360, East equal 90, South equal 180, West equal 270. This system is simpler and much superior to S 20 E for 160 or S 20 W for 200, for instance. It also avoids many misunderstandings and confusion.

Sample numbers are the collecting dates, for instance, 7177 is July 17, 1967, and A, B, C stand for the localities 1, 2, 3 on that date but years are indicated only for 1966, 67, 68, and 69. Cuts are labeled a, b, c. Old numbers are consecutive as, for instance, 47–344, or 1958, 1, 2, 3, etc. For sample localities see Figure 24.

With the extension of the study, methods described in 1947 (pp. 356–68) were modified and adapted to the larger project.

III

Methods and Procedures

Measurements were made on oölites, orientations of lineations and cleavages above and below the oölitic limestones, and slickensides, striations, and mineral orientations in the area between the east slope of the Blue Ridge and the west side of the Great Valley.

For oölites essentially the same procedure was used as described in 1947 (pp. 856–68):

1. Location of oölite beds from existing maps where possible.
2. Structural inventory: attitudes of bedding, cleavages, fold axes, lineations, intersections of planar structures, slickensides, growth of rods, and joints.
3. Determination of oöid distortion: longest, intermediate, and shortest axes and their orientations in the field with hand lens.
4. Marking of principal ellipsoid axes and cuts to be made with magic marker, colored pencil, or ink.
5. Specimen removal after labeling strike and dip on a smooth and even surface (Pl. 74).
6. Noting of strikes and dips of that surface in notebook and plotting of all data in an equal-area or stereo net in the field.
7. Preparation of sample for cutting and thin sectioning.

In the laboratory:

8. Making saw cuts in AC, AB, BC of the ellipsoids as accurately as possible.
9. Measurement of oöid dimensions under binocular with immersion oil and intense reflected light and study of surfaces; if need be cutting of additional surfaces which contain the principal axes.
10. Making thin sections of selected cuts and preserving orientations. Sections must be thicker than normal.
11. Study of sections petrographically and structurally. Dimensions should be measured and compared with saw-cut measurements.
12. Reorientation of data in equal-area net with field orientation. Net used as map with north at the top.
13. Listing of all data and plotting into maps, sections, diagrams, and graphs.

Finding and cutting the principal oöid axes for a sample is most important because a few degrees of deviation can change the values in fairly

7

high deformation ratios, and there is no way of detecting errors without controls. Oölite beds are easiest to locate in open fields where limestone surfaces are slightly weathered. Fresh artificial outcrops are almost hopeless, and in brush or forest limestones are covered with lichen and dirt.

In all, 310 oölite samples were collected: 409 saw cuts and 283 thin sections were measured for a total of 42,585 axial ratios distributed as follows: 418 AC cuts, 170 AB cuts, and 104 BC cuts. Some samples were not usable because deformation was either too intense or absent. Other samples turned out to be fragmental limestone or dolomite and not suitable. (For raw data, see appendix, p. 207).

The number of measurements per saw cut or slide depends on the intensity of deformation, kind of oölite, pre-deformation orientation patterns, and degree of recrystallization. If a constant value is attained with 35 or 50 measurements there is no need to continue. I have measured 100 dimensions in groups of 50 to control stability of maxima. Consistency of orientations was graphically determined as in Figure 25b.

Of 692 saw cuts and sections we determined 100 ratios in 283 and 35 ratios in 409. Controls were checked frequently. Many sections and cuts were measured two or more times.

We also measured saw cuts and thin sections in the same sample to determine possible differences. Kinds of particles must be separated because there is a vast difference between spherulites, mud pellets, layered oöids, sand grains, pre- and post-deformation alteration, and the association of matrix and oöids.

More than 25,000 field measurements were made and evaluated in diagrams, charts, and plots in maps.

Lineations, slickensides, striations, and *mineral growths* were treated elaborately because the existence of a systematic orientation in relation to oöids axes and lineations deserved more attention. A large number of diagrams were therefore prepared as follows:

The Skyline Drive was used as an operational base mainly in the study of slickensided surfaces because slickensiding and lineation seem related.

The Drive was divided into 19 subareas on the basis of the presence of Gneiss, Greenstone, or Chilhowee rocks. Large units of Greenstone were subdivided into small units. These areas are labeled in Figure 24, which also shows oölite sample localities. Table 9 lists area numbers, figure numbers, diagrams made, and rock types.

The orientation diagrams were plotted from field notebooks and the most densely populated girdle was determined. Finally, some diagrams were counted and contoured.

In Plates 1 and 2 I have presented the orientation of elements in space

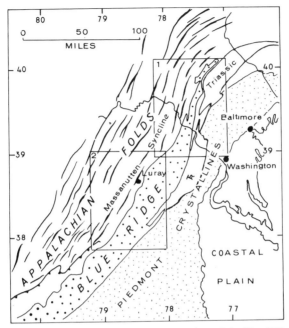

Figure 1. Areas and location of Plates 1 and 2 and the Blue Ridge.

and, in the oölites, an axial ratio of longest to shortest axis A/C. This is sufficient to trace deformation intensities and orientations of deformation planes and to design a regional model of the tectonic plan, but much important detail must be left to the geologist who can spend the time and wishes to study folding mechanism at selected points in relation to rock types and under conditions which vary widely.

Plates 3–61, 65, 66, and 82–87 are photomicrographs of thin sections. Plates 62–64 and 67–81 are photographs of samples.

IV

Description of Oölites

No two thin sections or saw cuts are alike because oölite beds vary, and because associations, compositions, matrix, primary structures, and secondary changes differ from layer to layer and from area to area.

In 1947 I distinguished spherulites, layered oöids and pellets, "pseudo-oölites," and detrital grains.

In Table 1 I have grouped oöids in order of decreasing usefulness for this study. Only the spherulites, layered oöids, detrital dolomite ("pseudo-oölites"), and quartz were originally spherical or almost spherical. In order of increasing ductility and decreasing preservation of shapes in a deformed oölite are detrital quartz and dolomite grains, spherulites and layered oöids, and mud pellets. Primary shapes are best preserved in detrital grains, fossil fragments, chert, and large mud lumps, and preservation is poorest in small mud pellets or in mud shells surrounding hard-core dolomite. Any of these types may be combined, and except for chert or dolomite oöids one seldom occurs alone.

To obtain a measure of the distribution of types in my samples I took at random the best 100 photomicrographs of oölite thin sections and counted the number of slides containing the various types. Results were as follows:

> spherulites 70, only spherulites 7
> layered oöids 63
> mud pellets 63, only mud pellets 3
> dolomite clusters 34, only dolomite clusters 1
> chert oöids 4
> pebbles 33
> fossil debris 21
> growth on mud 7

In 10 photomicrographs, distortion was too intense to distinguish types or to measure deformation. Good bedding was found in 16 photographs.

Deformation superimposed on the primary variability affects some types more than others. Thus dolomite clusters, sand grains, mud pellets, and chert are always recognizable, but spherulites and layered oöids may become unrecognizable.

11

Table 1. Oöid types used in this study

	Illustrations
1. Spherulites:	
a. Hard-core spherulites change to elliptical bodies partly by fracturing ("pie fractures").	Fig. 2a–d; Pls. 3–8
b. Spherulites with layered structures or with "rind." Layers inside or outside oöids, but not detached. No thickening of layers parallel to C.	Figs. 3, 5
c. Spherulites with detached "rind." Layers thicken parallel to C.	Fig. 4a–c
d. Spherulites with mud coating drawn out from spherulites parallel to C.	Fig. 4b and c; Pls. 23 and 24
2. Layered oöids:	
Distance between layers increases parallel to C. Some layers may be spherulitic or radial growths.	Figs. 5, 9; Pls. 9–14
3. Mud pellets:	
Since original shapes were not spherical, mud pellets are rarely usable except to indicate a lineation direction.	Pls. 15, 16, 35, 36
4. Dolomite clusters and single grains:	
Commonly undeformed even in highly deformed oölites. Deformation by fracturing, boudinage, or disintegration to individual grains.	Fig. 10; Pls. 21–24
5. Chert oöids:	
Chertified oöids are preserved in chert nodules.	Pls. 25, 26

To avoid recitation of lengthy details, the descriptions have been placed opposite the photomicrographs on facing pages. Only general observations are here presented. Additional descriptions are incorporated in the discussions of oöid assemblages, distribution of ratios, and orientations.

The wide variety of spherulites and their associations with other types are well illustrated on Plates 3–8. Layered oöids are shown in Plates 9–14. There are many gradations, because spherulites can have concentric rings, and layered oöids have spherulitic centers.

Pellets (Pls. 15–20) are very abundant in all oölitic limestones, and few samples are without any. Unfortunately, the original shape of pellets is so variable that they can hardly be used in quantitative strain analysis. They do form cores of spherulites and are then transformed into much more useful strain markers, because the spherulitic overgrowth may render the entire oöid spherical. Even elongate fossil fragments may become nearly spherical through fibrous overgrowth.

Since pellets are ductile and offer little resistance to deformation they furnish an indication of the behavior of the matrix when compared with spherulites or dolomite clusters.

Even if pellets cannot be relied upon as quantitative strain markers they are useful indicators of extension directions. When deformation is

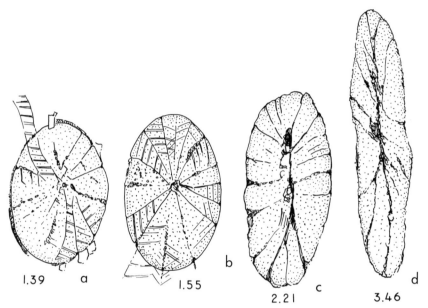

Figure 2. Four spherulites with increasing A/C ratios. Wedges with parallel calcite cleavage (b) have rotated at the ends and become narrower and longer (d) parallel to extension direction. In the center, wedges have become wider and shorter normal to A and were pulled apart. Center of spherulite is an elongated zone of debris (d). Locations: (a) 71 c, Shepherdstown, W.Va.; (b) 713D, west of Waynesboro, Va.; (c) 711 north of Shepherdstown; (d) 7177c, south of Luray, Va.

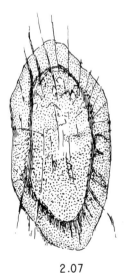

2.07

Figure 3. Elongated spherulite with undetached mantle and longitudinal fractures which suggest cleavage (S_1), south of Luray, Va.

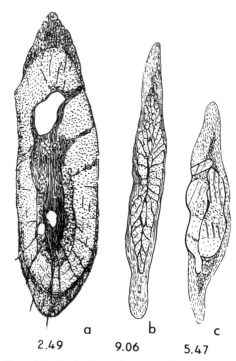

a   b   c

2.49   9.06   5.47

Figure 4. Spherulites with detached mantles: (a) usable spherulite; (b) extreme deformation, value questionable; (c) crystallized center has been stripped. Mantle provides good data on extension directions but not on strain quantity. 717, south of Luray, Va.

intense large, odd-shaped pellets become elongated also. Where spherulitic growth occurs on pellets, ellipsoidal shapes can be used unless centers are long fragments which are included due to size or pronounced odd shape as in Plate 16 or Plates 17 and 18.

Pellet sizes are important because small pellets are readily oriented, large ones are not (Pls. 16 and 18). Large pellets may be composed of small ones or may contain spherulitic oöids (Pls. 16 and 18). The spherulites may then be elliptical within the large pellet with ratios equal to outside ratios.

Dolomite (Pls. 21–24) is common in many oölites as single grains, clusters, or in replacements. Single grains and clusters appear in centers of oöids or as "pebbles" or balls or scattered throughout the sample. Dolomite layers are also common.

Dolomite is of special interest because it resists deformation in an environment of calcite and mud. If it is deformed at all it is broken. The problem was mentioned in 1947, and dolomite balls were pictured (E. Cloos 1947, Pl. 2, dolomite grains in Fig. 1 of Pl. 3 and clusters in the center of oöids in Fig. 3 of Pl. 5). Several drawings of dolomite are shown on Plates 7 and 8 of that paper.

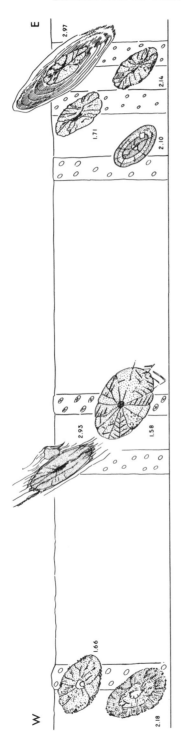

SECTION WEST OF WAYNESBORO VA.

Figure 5. Section across Conococheague Formation west of Waynesboro, Va. Oölite beds about 500 feet apart. Oöid drawings vastly enlarged to show types of oöids. Ratios fluctuate with types. Directions are consistent.

For a more detailed discussion of deformation of components see chapter V, "Ratios and Species of Oölites."

Chert nodules are common in oölitic limestones and are commonly oölites. They were not systematically collected but are valuable indicators of deformation, especially of early stages. Chert seems to have formed after deformation began, because the oöids are elliptical in the AC plane, but the ratio is much less than that of the surrounding matrix and oöids in it. In addition, directions of axes have changed, and rotation seems to have occurred. In the two pictured samples (Pls. 25, 26) rotation is counter-clockwise with the top to the west. The distance between the samples is 54 miles along strike. Both samples are in the east belt with high ratios, distinct cleavage, and lineation adjacent to the Blue Ridge upland to the east. It may well be possible to determine rotation systematically by using chert nodules, because they include a fairly long period of deformational history.

Pebbles (Pls. 27–30) are large clusters of varied composition. They can be mud lumps with quartz grains, dolomite, or spherulite, or they are small-grained pellets in a spherulitic oölite or almost any combination. They are useful in timing deformation stages and in analysis of intensity of deformation and its effect on primary shapes of components. They also provide data on mineral constituents which are preserved within a pebble. Some pebbles are only baked-together oölite with a somewhat different matrix and less recrystallization. Others include different oöids, mud pellets, or detrital material that is otherwise not present in the host rock, such as quartz or dolomite grains. (Pebbles occur also in Pls. 8, 14, 16, 18, and 19.)

Fossil fragments (Pls. 31, 32) occur in most samples. Small ones are incorporated in oöids as centers, larger ones may be slabby and show growth on their surfaces. Fossils are not as useful as oölites, where oölites are as abundant as in the Great Valley. We have found very few, possibly because of the intensity of deformation along the foothills of the Blue Ridge. (See also Pls. 18 and 29.)

Bedding can be seen in many of the photomicrographs. It can be a lithologic change, a layer of mud, insolubles, or stylolites and is very useful as a direction marker or for control of orientations. Two very distinct bedding planes are shown in Plates 33 and 34 and others are seen elsewhere (Pls. 5, 8, 12, 18, and 24).

When an oölite becomes extremely deformed it may be impossible to recognize it as oölitic (Pls. 35, 36). An occasional remnant spherulite or pellet may be preserved, and AC sections are usually hopeless. Sections with smaller ratios, for instance AB, however, show easily recognizable oölite structures, layered oöids, spherulites, and even pellets (Pl. 62). South of Luray, on U.S. 340 on the road to Antioch church, are outcrops with excellent oölite in fresh and weathered surfaces but nearly unrecognizable in AC surfaces. Such examples are at the limit of what can be reccognized as an oölite or measured, and more deformation results in a streaky schistose, and very lineated, limestone. There are many such examples in the foothills of the Blue Ridge between Lexington and the Potomac River.

V

Ratios and Species of Oölites

If several kinds of oöids occur together in an oölite they may differ widely in registering deformation intensity (E. Cloos 1947, p. 872). In "mapping" deformation intensities by means of A/C ratios it is therefore essential to separate different species and to compare only similar ones.

The following nine combinations of oöids which occur together most frequently within an oölite were measured separately: (1) spherulites and mud pellets; (2) spherulites, small and large pellets, spherulites in mud matrix; (3) spherulites and small pellets; (4) different sizes of spherulites; (5) mantled spherulites and mud pellets; (6) oöids inside and outside oölitic pebbles; (7) spherulites within and outside pebbles; (8) chertified oöids and mud pellets with spherulites; and (9) "pseudoölites" and undeformed crystal aggregates in mud or calcite matrix.

The regional distribution of ratios will be dealt with below.

Spherulites and mud pellets

These were measured separately in the AC section of one sample (Pl. 8). Average ratio for 100 spherulites is 1.68. Average long axis is 0.54 mm.

Mud pellets are much less evenly shaped and less well oriented than the spherulites. The large pellets are randomly oriented, and smaller ones are roughly aligned; the smallest ones are elliptical and oriented. Average long axis of pellets is 0.75 mm. Ratio is 2.18. This ratio may be misleading because the original shape of the mud pellets is not known. Their irregular shape had to be overwhelmed before they could become elliptical. Spherulites began as spheres, whereas mud pellets had to become more or less isometric before they could be extended into elliptical bodies (Elliott 1970). The two large slabby pellets of Plate 8 would have to be shortened to one-half of their lengths and thickened in the direction of spherulite extension. This would make them squares. At that stage the spherulite ratio would be 2. From then on the pellets could be extended but would lag behind the spherulites, provided the entire rock were to act homogeneously. The same principle is illustrated in Plate 16.

19

Spherulites, small and large mud pellets, and spherulites in mud matrix

This combination was measured in order to determine the deformation of spherulites in mud and in crystalline matrix which may have been mud before crystallization, and the deformation of small and large mud pellets, in order to compare them with spherulite deformation. The two sizes chosen were: equal or smaller than the spherulite average, and larger than the spherulite average (Fig. 6; Pl 19).

In Figure 6 sharply defined spherulites are embedded either in calcite or in mud. They have sharp boundaries with occasional concentric layering and a cross in polarized light. Some of the spherulites have small mud pellets or crystals as centers. They grade into large odd-shaped growths on large mud pellets or pebble-like units.

The A/C ratio for 100 spherulites embedded in mud is 1.52. Outside mud in crystalline matrix, the ratio for 100 spherulites is 1.48. This difference suggests that crystallized matrix suppressed distortion of spherulites whereas mud facilitated it.

Ratio A/C for the small pellets is 2.43 and for the larger ones is 1.84. Obviously, the smaller pellets are more readily oriented and may have been more nearly spherical originally. The larger ones are irregular, poorly aligned, and partly pebble-like. Some large slabs parallel to bedding were not measured. Both pellet groups show higher ratios than the spherulites.

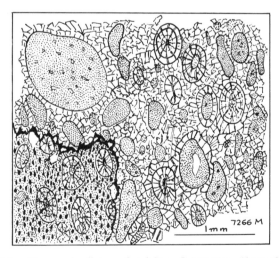

Figure 6. Spherulites, mud pellets, and stylolites. Orientation of layered spherulites and small pellets is constant. Large pellet with apron is unoriented. Stylolite cuts across orientation. 7266Ma, 5 miles east of Harrisonburg, Va.

Table 2. Ratios and species, summary of data

		A/C
Spherulites and mud pellets:		
(Pl. 8)	Spherulites A/C	1.68
Sample 7226G		
Area 2	Pellets	2.28
Spherulites, small and large pellets and spherulites:	Spherulites in mud	1.52
in mud matrix. (Pl. 19)	Spherulites in crystalline matrix	1.48
Sample 7266M	Small pellets	2.43
Area 3	Large pellets	1.84
Spherulites and small pellets:		
(Pl. 37)		
Sample 8176Ga	Spherulites	1.53
Area 3	Small pellets	2.30
Sizes of Spherulites (Pl. 3):		
Sample 7256Ca	Large spherulites	1.18
Area 2	Small spherulites	1.18
Spherulites, mantles and pellets:		
(Pl. 13)		
Sample 713Gb	Mantled oöids	2.40
Area 3	Spherulite cores	1.79
	Spherulite naked	1.81
	Pellets	4.35
Strain inside and outside pebbles:		
(Pls. 8 and 13)		
(Pls. 18 and 19)	Layered oöids	2.91 ⎫
(Pls. 27–30, and 38)	Layered oöids	2.78 ⎪ in Pl. 38
(Pls. 8, 18, 27–29: Area 2).	Pellets inside pebble	3.63 ⎬
(Pls. 13, 19, 30, 38: Area 3).	Pellets outside pebble	3.74 ⎭
Spherulites inside and outside:		
Pebbles (Pl. 39)	Spherulites inside	2.19
Sample 813A4	Spherulites outside	2.00
Area 3		
Spherulites and pellets in pebble:		
(Pl. 30)	Spherulites inside	2.35
Sample 813A2	Spherulites outside	2.27
Area 3	Pellets inside	3.08
	Pellets outside	3.15
Chert (Pls. 25 and 26)	Chert oöids in nodule	1.13
Sample 7216Ba	Pellets in contact zone	1.26
Area 3	Pellets away from nodule	4.88

Table 2.—*Continued*

Chertification partial:		
(Pl. 40)		
Sample 728A	Spherulites in matrix	2.29
Area 3	Chert oöids	2.14
(Pl. 41)	Chert in nodule	1.51
Sample 5814B	Chert intermediate zone	1.74
Area 3	Spherulites away from nodule	2.00
	Pellets	3.76

Mud is more readily deformed than spherulites, but large pellets with axial ratios such as those of Plate 16 are not readily oriented, and it takes more deformation to reach the ratios of the small pellets (Elliott 1970).

Spherulites and small mud pellets

The relationship between spherulites and small mud pellets is further illustrated in Figure 7, a drawing of an AC section (Plate 37). This oölite consists of spherulites, layered spherulites, small mud pellets, and crystalline matrix. The spherulites are elliptical, well defined, and surrounded by small growth aprons.

The mud pellets are like a school of small fish that surrounds a large one. Their boundaries are fuzzy, and their texture is very fine-grained and granular. Very small carbonate grains can be identified in a muddy matrix. Grain size is much smaller than that of the matrix. The original shape of the pellets is not preserved. Their even distribution throughout

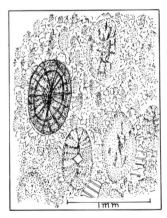

Figure 7. Spherulites, large and small pellets in matrix of crystalline calcite. A/C ratios for 100 spherulites 1.53; for small pellets 2.30. Orientation is not deflected by large spherulites (see also Plate 37). 8176Ga, U.S. 60, Buena Vista-Lexington, Va.

the thin section and also their rather even size suggest pellets and not a large mass of disintegrating mud.

A/C ratio of 100 spherulites is 1.53; the ratio of the small mud pellets is 2.30.

The small mud pellets may be useful strain gauges because they are fairly evenly distributed around the much larger spherulites. They may have preserved a slightly more mobile medium in which the spherulites were deformed.

If the spherulites had resisted deformation as much as the dolomite clusters, the matrix and mud pellets would flow around the spherulites in "streams" of pellets as mica or cleavage flows around feldspar in a schist. The spherulites would be "Augen." This is not the case; deviations of the mud pellet "trains" are very faint. The orientation of mud pellets is almost entirely in the direction of the long spherulite axes and continues through them to the other side.

Spherulite sizes

Different sizes of spherulites were compared in several samples, but appreciable differences of ratios were not found. Different sizes shown in photomicrographs (Pls. 3–7 and 9) are due either to different species, as in Plates 3 and 4, or to sections which are not center sections but are high or low ones. Plates 3 and 4 show two oöid species: large, dark, brown, well-layered spherulites and light ones half their size.

Ratios for both 100 large and 100 small spherulites are 1.18, and de-

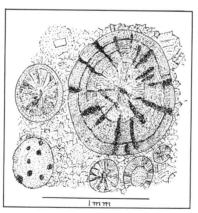

7256Ca

Figure 8. Radial structures in spherulite reaches periphery in center cuts. Radii appear as dark dots in high or low cuts. They are black lines from center to periphery of unknown origin and very common in all spherulites (see also Pls. 43, 44). North River, R.R. track south of Harrisonburg, Va.

formation intensity is very consistent across the slide, in spite of the heterogeneity of the sample.

Spherulites, mantles, and mud pellets

Spherulites, mantles, and mud pellets were compared in the AC section of sample 713Gb (Fig. 9; Pl. 13).

A large oöid consists of a normal spherulite and a dark, finely laminated, granular mantle which is pulled away from the core in the extension direction, fraying at the ends and becoming indistinct.

The A/C ratio for mantled oöids is 2.40, for spherulite cores 1.79, for naked unmantled spherulites 1.81, and for mud pellets 4.35. Fifty measurements were made for each variety, and an extra 100 for the large mantled oöids, whose ratio is 2.40 (Fig. 9). The spherulite ratio is the lowest and is the same for mantled or naked spherulites.

Oöids inside and outside oölitic pebbles

Oölitic pebbles consist of oöids in a mud matrix. Outside the pebbles the spherulites are in a crystalline matrix. The pebbles may contain only

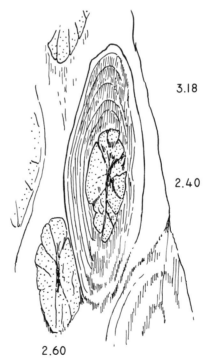

3.18

2.40

2.60

Figure 9. Spherulite with mantle detached in elongation direction. Mantle ratio (3.18) higher than spherulite ratio (2.40–2.60). Detached mantle shows cleavage. 713Gb, west of Waynesboro, Va.

small mud pellets in a muddy matrix, and some pebbles are only large mud lumps with carbonate or quartz grains.

Several samples were chosen for comparison of A/C ratios inside and outside the pebbles. Sample 886G2 (Pl. 38) shows the following average A/C ratios in thin section:

100 layered oöids	2.91
50 layered oöids	2.78
50 mud pellets outside pebble	3.74
50 pellets inside pebble	3.63

In order to compare all kinds of mud pellets the ratios of 50 mud-aggregate centers within layered oöids were measured; the ratio is higher than that of the layered oöid, but not as high as the "free" mud pellets.

The pebble seems more deformed than the rest of the slide. This is due to the high concentration of pellets in the pebble, where high ratios are not blurred by normal layered oöids, and fairly large spaces between that are filled with clear calcite matrix. The ratio for pellets in the pebble is 3.63; for the layered oöids it is 2.78. However, if comparable oöids are compared the difference is slight and may well be within the limit of error: mud pellets inside are 3.63 and outside 3.74. Since pellet boundaries are not really distinct, and pellets are granular aggregates without sharp boundaries, the ratios are the same. The pebbles have participated in the deformation along with the mud pellets scattered in the matrix. Regrettably there are no layered oöids in the pebble for comparison with the same species outside.

A direct comparison of spherulites inside and outside pebbles is rarely possible because pebbles with spherulites in them are not abundant; only a few spherulites are embedded in a muddy matrix.

One good example is specimen 813A4 (Pl. 39).

The A/C ratio of spherulites outside the pebble is 2:00, and inside the pebble it is 2.19. This small difference may be due to the difficulty of determining the boundaries of spherulites that are intensely distorted.

Another example was taken from sample 813A2 (Pl. 30).

A/C ratios are as follows: spherulites outside, 2.27; spherulites in pebbles, 2.35; mud pellets outside, 3.15; mud pellets in pebbles 3.08.

Thus for comparable materials the ratios within and outside the pebble are essentially the same. The difference between 2.27 and 2.35 is insignificant. Mud-pellet ratios are 3.15 and 3.08, also very close, but higher for the mud pellets outside than for those in the pebbles.

More significant is the difference between spherulites and mud pellets, as in other slides, mud-pellet ratios are higher than spherulite ratios.

Chertified oöids and rind pellets with spherulites

Oölitic chert nodules are quite common, and several examples were analyzed. The nodules are embedded in a dense, black, muddy matrix which is streaky with elongated mud pellets and spherulite remnants that "flow" around the nodules (Pls. 25, 26, 40).

In Plate 25 the deformation is uneven, crystalline centers remain undeformed, and only the mantle is drawn away. Where the deformation is unhampered by the nodule, clusters of elongate carbonate crystals or broken aggregates are scattered throughout the streaky mud. Chertification was early when deformation was still slight, but sufficient to turn a slightly random ellipticity into an orientation that affected the area which is now chert. Chertification does not stop abruptly at the contact; some carbonate centers outside the nodule are also chert. As deformation continued the chert nodule froze and rotated in the mud-oölite matrix. The A/C ratios are as follows: within the chert nodule, 1.13; in the contact zone but not chert, 1.26; away from the contact, 4.88.

Partial chertification in a matrix of carbonate but not in a well-defined chert nodule is shown on Plate 41. In the matrix and surrounded by mud streaks and granular calcite 30 spherulites could be measured. They are well defined, partly layered with an outer shell and granular centers and a dark cross in polarized light. The ratios are: spherulites in matrix, 2.29; chert oöids, 2.14. Mud pellets cannot be measured. The difference between 2.14 and 2.29 is possibly significant but is rather close. At that ratio the matrix spherulites are not as sharply defined as the chert oöids, and especially the A termini become hard to define. Although only 30 could be used many spherulite fragments can be seen; most of the matrix may indeed be fragmented spherulite.

For chert oöids the ratio of 2.14 is the highest measured. The stage at which chertification froze the oöids to their present shape must precede the additional deformation suffered by the matrix, because the chert oöids are still very sharp and clearly outlined with light haloes of quartz fibers. The matrix is streaky and was carried beyond the chert stage.

The gradation between deformed spherulites and arrested deformation is well shown in sample 5814Bb (Pl. 41). Ratios were determined for the following oöids: chertified oöids in chert nodule, 1.51; spherulites in area with chert forming as matrix between carbonate spherulites, 1.74; spherulites at least five grains away from chert, 2.00; and mud pellets, 3.76. Orientation is undeflected across the entire slide.

The slides suggest that chertification froze some of the spherulites in an early stage of deformation. Deformation continued in the matrix of an adjacent area, and in the embedding oölite it continued still farther, but there is no chertification. The sequence may have been as follows:

chertification completed for chert oöids at ratio	1.51
partial chertification but additional deformation to ratio	1.74
no chertification, spherulites deformed in embedding medium to	2.00
continuing distortion between spherulites affects mud only and terminates at mud-pellet ratio	3.76

However, the original shape of the mud pellets is unknown and was not necessarily spherical.

Dolomite clusters or pebbles

In 1947 I called attention to the "pseudoölites" of carbonate clusters (1947, Pls. 2, 3; Pl. 7, Figs. 1, 10; Pl. 8, Figs. 7, 8, 9, 11) and to the fact that these "pebbles" are undeformed in an otherwise strongly strained area. This can be seen in almost any sample. The carbonate aggregates, mostly dolomite, are even less strained than are the chert oöids. They were spherical early and remained so; where deformation was extreme they fell apart into rows of carbonate crystals, single crystals, clusters, or even augen of dolomite. But far more commonly they remained spherical in a highly strained environment of mud streaks, mud pellets, and spherulites.

There are suggestions that the "pebbles" are replacements of calcite spherulites at a pre-deformation stage. Some show a crude spherulite cross which is then a successive extinction of adjacent crystals. The existence of a cross depends on the number of crystals. If there are many small ones a cross may show in the orientation of the extinction. In a sphere with only one or two crystals, the cross has been eliminated. The cross can even be suggested by four carbonate crystals if two grains in opposite quadrants extinguish simultaneously.

The second replacement feature is a mud center or single crystal in the center surrounded by a ring of several or even only one crystal. Centers are seen in several pebbles of Plate 23 and in Figure 10. The concentric structure probably demands an oölite origin.

The pre-deformation shape and origin are proved by the lack of distortion, the random ellipticity in an otherwise well-aligned environment, the rind which envelops the pebble, and the flow of mud around the circular pebbles (Fig. 10). Many of the pebbles show growths of carbonate fibers in pressure shadows parallel to C.

Dolomite clusters are unsuited as strain gauges because they do not picture any deformation. They do indicate that the difference between the hardness of dolomite and calcite was sufficient to protect them. This

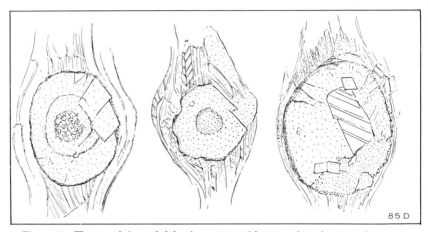

Figure 10. Three undeformed dolomite centers with suggestion of concentric structures. Large crystals are surrounded by muddy layers that flow around them. Some fibrous calcite grows at ends. If centers are replacements they must be pre-tectonic, preserving an original spherical shape. 85D, south of Berryville.

mutual relationship of fabric partners (*Einfluss der Gefügegenossen*) was stressed years ago by Bruno Sander (1930).

The very beginning of strain effects is seen in low-ratio deformation (Pl. 42). Dolomite clusters in a bed between limestones show less deformation than the nearby oölites. Fractures are, as usual, more abundant in the dolomite; boudinage occurs, and this difference in deformation is also mirrored in the relationship between spherulites and dolomite clusters.

The contrast of strain between spherulites and dolomite "pebbles" is illustrated in Plate 21. The difference between readily deformed mud or layered oöids and unstrained dolomite cores is shown in Plate 22. Small "pebbles" are scattered throughout the slide, and many consist of only one or two large grains. There are also mud oöids, some well-layered mud, and calcite, with centers that are larger than the "pebbles"; dolomite also occurs as single crystals or clusters of crystals. These centers have dark rinds and are undeformed and almost spherical. Around the center is black granular mud with very small carbonate grains. They are not spherulites and show no suggestion of a cross. The solid black lumps are mud oöids which were cut above or below the center by the thin section. The matrix is calcite and small mud wisps. The ratio for 100 mud oöids is 2.14.

The most extreme difference between dolomite clusters and mantled spherulites is shown in Plate 23. Spherulites have granular cores and once-concentric layering.

Three sets of measurements were made on this sample:

1) The mantle with spherulite centers; 200 grains with an average A/C ratio of 4.57, which is minimal.
2) Spherulites in mantled oöids, ratio 4.00.
3) Dolomite cores, ratio 1.29.

All measurements are approximate due to the intensity of distortion, the loss of material, and the fragmentation of hard cores or loss of coherence. It is noteworthy that it was possible to find 40 complete pebbles, well rounded, with their rind of mud still intact and yet slightly elliptical (A/C ratio 1.29). The orientation of the long axes of these ellipses is not consistently parallel with spherulite axes. This may suggest a primary non-tectonic ellipticity. In that case the ratio of 1.29 is not an indication of strain.

The difference between the ratio of 1.29 and 4.00 or 4.57 indicates that deformation was not beyond the strength of dolomite clusters or even single crystals as long as these were in a matrix of mud or calcite.

Matrix and oöids

What the matrix was at the time of deformation is not known, but it is now crystallized calcite that shows little deformation. It may therefore be largely post-tectonic or late tectonic. Some of the matrix fills spaces by growth on deformed oöids, particularly in the direction of oöid extension.

In a carbonate matrix quartz grains grow into euhedral crystals across boundaries and across deformation co-ordinates (Pls. 20, 27); in quartz veins and similar environments, however, the quartz is generally intensely deformed, fractured, and brecciated (Pls. 82–85).

The question arises: Did spherulites and matrix deform homogeneously together to make oöids representative of the strain of the rock or at least of the relative strain difference between areas of considerable strain gradients?

Assuming that the matrix were soft and the oöids harder, they would be crowded together, and contact effects would be more common. Plate 9 shows unusually crowded but barely touching oöids, with little effect at the contacts. Many oöids are single, surrounded by matrix and yet deformed at the same ratio as all others. In Plate 11 deformation is much more intense, oöid sections are elliptical, and feeble contact effects are present. Plate 10 shows widely spaced oöids, elliptical in a matrix of mud and calcite. Oöid sections cut through the center show that the oöids rarely touch but are supported by the matrix.

Plates 3 and 4 show several kinds of oöids: small spherulites and dark ones twice as large, with closely spaced laminae and spherulite centers. The matrix consists of debris, fragments, fossil fragments, irregularly shaped growths, small mud pellets, and clear crystalline calcite. Radial structure prevails in all oöids. Small and large oöids have very similar ratios, and elongation is also uniform and easily recognized, even at a low ratio of about 1.20. Stresses must have been exerted rather uniformly across the entire slide and not from oöid to oöid, but through the matrix. If the matrix had not transmitted stresses across the slide, oöids would have been pushed together and touched each other. Where they touched, interpenetration and the effect of pressure would be visible. In spite of the fact that oöids rarely touch each other, deformation is registered equally in widely spaced and closely packed oölites, suggesting that the matrix did indeed transmit stresses across a slide or a sample. Where differences exist, as between dolomite clusters or unyielding bodies and mud pellets or matrix, the matrix flows around these bodies, and measureable differences are obvious. I conclude that if there are no measurable differences, oöids and matrix can be assumed to have been deformed together, and the oöids represent very closely the strain of the entire sample.

VI

Micromechanics of Oöid Deformation

The deformation of a nearly spherical oöid into a triaxial ellipsoid is not a simple process. Surprisingly little has been published on the subject, and the distortion is barely described except in terms of axial ratios and the resulting shapes which have been used as strain ellipsoids. However, closer study of the oöids is well worth the effort, because different oöid species behave differently under stress, and, accordingly, they furnish data on the process of deformation that may be significant in strain analyses and in the use of oöids as strain ellipsoids.

The several species of oöids vary widely in their reaction to deformation. Most mobile, "weakest," and most readily deformed are mud pellets, but their original shapes are not spherical. They are valuable as indicators of directions where deformation is intense, but as strain ellipsoids they are unreliable. At the other extreme, the undeformed dolomite pebbles, sand grains, or aggregates show little or no distortion and cannot be used as strain indicators. Between the two extremes are oöids with concentric layering, spherulites, and combinations of these two. Most useful and most common are the spherulites, whose structures were nearly identical before deformation and whose patterns of distortion are comparable.

The deformation of near-spherical or spherical bodies in tectonics has been discussed recently by Stauffer (1970), whose illustrations of a large variety of deformed bodies and textures are outstanding and strikingly similar to those here described. Stauffer also stresses the processes that are involved in oöid distortion and the gradation from plastic deformation at one end and brecciation at the other with crystallization and displacement of particles between the two. He distinguishes brittle behavior, intracrystalline dislocations, intergranular dislocations, and recrystallization (Stauffer 1970, Table 1, p. 500). The deformation of spherulites combines all these at various stages.

Spherulite deformation

A spherulite is a "spherical brush" with fibrous growth, radiating from a center. Concentric layering is common, mostly by color banding or

31

lamination (Pls. 43–46). Fibers continue across banding and make the well-known extinction across in polarized light. Deformation involves rotation of fibers, adjustments between fibers, crystallization of fibers into larger wedges, changes of the core, and changes in the surrounding matrix and especially in the calcite touching the spherulites. If the spherulite is layered, the layers may behave like incompetent layers in folds. They may thicken, detach from the lower beds, and leave room for growth in voids. Beyond a given point, outer layers migrate away from the spherulite, and the ends of oöids become frayed (Pls. 11–14). In progressing deformation, rotating fibers of spherulites are faulted, the surface of spherulites is offset (Pls. 50–55), and the rotated units, enlarged as bundles of fibers, are welded into one crystal. Between wedges appears fine-grained calcite or mud (Pls. 47, 48, 50–53).

In addition to a complex primary structure different portions of a spherulite are strained differently. As illustrated by Ramsay (1967, p. 119), when a sphere is changed to an ellipsoid the different parts of the sphere experience different conditions, which are beautifully shown in deformed spherulites. The A axis is extended, C is reduced. In a wedge between A and C reduction is followed by extension. It is also an area of considerable rotation. As deformation progresses fractures appear, and the fracture patterns differ in the AC, AB, and BC surfaces. Finally, fractures may coalesce across a slide or sample, and cleavage also appears (Pls. 50, 51, 54, 55). It is not clear how much, if any, ductile flow is involved in the deformation because only low-ratio deformation lacks fractures, offsets, rotation, or granulation (Pls. 43–48).

Cores of spherulites can be nonspherical and can influence the deformation history even if an elongated core is covered by spherulitic growth which masks the eccentricity of the core (Pl. 43). A spherulite is thus a complicated structure in which different portions of a sphere undergo a variety of changes, as follows:

Growths: (1) as beards at the ends parallel to C; (2) between fibers or bundles; (3) of fibers into larger one-crystal wedges; (4) of "aprons" beyond the spherulite surface; (5) in fractures.

Mechanical Changes: (1) rotation of bundles or wedges; (2) slippage between rotating units; (3) rotation of planes; (4) jointing; (5) faulting; (6) granulation.

The process is complicated in nonrotational pure shear, but becomes intricate when the entire system is "rolled." Then differences in the four quadrants indicate rotational deformation and the sense of rotation.

The mechanism of spherulite deformation is schematically summarized in Figure 11 and can be traced on Plates 43 to 51, and 54, 55. Deformation begins probably with a distortion of the oöids along fibers and wedges. If the wedges are small or single fibers, slight displacements of the surface are not visible, and the displaced wedges are so small that the surfaces appear continuous. Primary radial structures resembling wheel spokes provide favorable surfaces which limit wedges and permit displacements.

Figure 12 is a drawing of extremely elongated spherulites. It is not clear when recrystallization of fibers into larger wedges takes place. Wedges are recognizable only when distortion results in an A/C ratio of 1.4 and when rotating wedge boundaries are compressing the pie-shaped wedges. Then twinning permits recognition of larger crystals. With more intense deformation the spherulites break into fragments.

Fractures in oölites

Oölites are intensely fractured, and a thorough study of fractures may well yield valuable results, particularly in relation to built-in strain markers like oölites. Some microfracture patterns are obviously related to oöid deformation, and they are here described because they are part of the deformation pattern. Fractures are not readily seen in all thin sections or all cuts. Some radial fractures are best seen in center cuts, and they may not be visible in cuts above or below the center; others are seen only in high or low cuts.

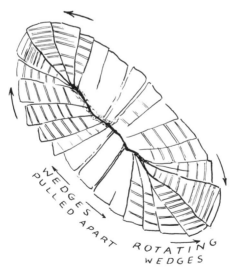

Figure 11. Schematic presentation of spherulite deformation. The original spherulite consists of fibers radiating from a center. The fibers consolidate and recrystallize into one-crystal wedges with a common calcite cleavage. As deformation continues the wedges rotate at the ends and are compressed. In the center they are shortened and pulled apart into a long zone of debris.

Some fractures are not visible in AB cuts but are very distinct in AC cuts. BC fractures are not seen in BC cuts, are best seen in AB cuts, and are most distinct in high or low cuts.

The several kinds of fractures are illustrated as follows:

Fractures limited to oöids
  Radial fractures: Pls. 49, 51–55, 58, 59.
  Unoriented fractures in oöids: Pls. 56–59.
  Longitudinal fractures in oöids: Pls. 54, 55.
  Fractures filled with carbonate limited to oöids: Pls. 51, 54, 55.
  Rotated tension fractures filled with carbonate: Pl. 51.
Fractures across thin section or sample (see also Table 3)
  Shear fracture patterns: Pls. 7, 11, 47–49, 52–55, 58, 59.
  Fractures filled with carbonate: Fig. 13, Pls. 8, 18, 21, 31, 34, 50.

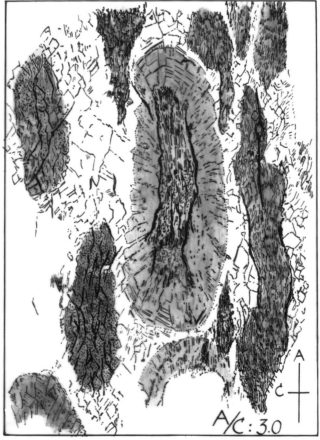

Figure 12. Spherulites as in Figure 11. One large elongated center cut with compressed ends. High or low cuts show longitudinal fractures (cleavage ?). 47–344, south of Shippensburg, Pa.

The radial primary fabric of spherulites must influence the formation of radial fractures, just as the fabric of wood facilitates splitting parallel with the grain. The radial fabric should favor radial fractures or fractures across spherulites through the center. If the oöids are closely packed, fractures in oöids can coalesce into cross-cutting patterns which include the entire thin section.

Since spherulites are fibrous brushes fractures form across the center parallel with the fibers more readily than across fibers above or below the center. They may also develop from the center outward. As soon as fibers are recrystallized into wedge-shaped crystals, the wedges between crystals become sensitive surfaces, and fractures appear.

Plates 43 to 59 show individual spherulites with increasing deformation ratios between 1.10 and more than 3.0.

The primary radial fabric of spherulites facilitates fracturing between bundles of fibers or between recrystallized groups of fibers. Spherulites are more readily fractured than is the randomly oriented crystalline matrix. With low deformation, fractures are radial surfaces which be-

Table 3. Microfracture patterns in oölites.
(List of angles between fractures (center angles) and angles between long spherulite axis A and shears, mostly symmetrical with respect to long axis A.)

Ratio A/C	Left	Angles center	Right	Area	Plates	Samples
1.31	35	111	34	2	49	420
1.33	34	110	36	3		364
1.62	70	28	62	3	50	371
	45			3		372
2.00	37	109	34	3		381
	30	120	30	3		389
1.4	43	115	22	3		403C
2.06	26	112	42	2		422
1.34	50	75	55	2	58	58–16
1.19	40	91	49	2		58–18
1.80	50	70	60	3	7	812Hb
2.72	17			3		716Bb
1.60	22			3		86A4a
1.82	27	95	58	3		812I3
1.62	12	118	50	3		812E3a
2.1	40	108	32	3		812Ba
1.8	37	114	29	3		466A
2.5	32	115	33	3		466C
2.42	16	130	35	3		7206B
1.56	14	104	62	3	11	7256F
1.61	63	62	55	2		6157Eb

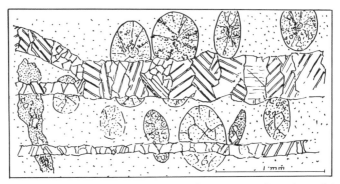

Figure 13. Calcite-filled tension fractures normal to extension. Secondary calcite in fracture coarser than in host oölite. Crystals grow normal to fracture wall. 47–419, 1 mile east of North Mountain.

come faults by rotation of fiber bundles or wedges. They play an important role in oöid distortion and become more obvious as deformation progresses (Pl. 51). Fractures coalesce to patterns which transect thin sections or whole samples (Pls. 7, 11, and 58). Similar patterns were measured in 18 additional AC sections of which 15 are in the east belt (Area 3) and 3 the western and central belts (Table 3). The last fractures to form are calcite-filled veins, mostly tension joints normal to A (Fig. 13).

Fractures begin to form at low ratios and grow more prominent as ratios increase. The question arises whether oöids and possibly other strain markers were ductile bodies that were distorted homogeneously with the surrounding rock. This seems probable since mud streaks and matrix do not normally flow around spherulites, but do flow around dolomite clusters and grains that remained undistorted. Table 3 suggests that most values are of the same order of magnitude. The average angle between two directions is larger than 100 degrees. Only a few fracture sets are highly asymmetrical with reference to the A axis.

The role of spherulite fabric in fracturing is well emphasized by the fact that very muddy slides with few spherulites show no fractures. At relatively low-intensity deformation, spherulites may be necessary for the formation of fractures, except for late calcite veins which form everywhere and are of much larger orders of magnitude.

VII

Regional Distribution of Deformation Intensities

Deformation intensities are expressed as A/C ratios and vary widely between oöid species, between areas, within areas, and even within locations.

The interrelation between ratios and oöid species has been presented in two ways: The species were listed (Table 1) and described, and their coexistence during deformation was analyzed. Their behavior becomes still more obvious if A/C ratios are plotted for spherulites, mantled oöids, and pellets for the three oölite belts.

Figure 14 shows the A and C values for areas one and two, Figure 15 shows them for Area 3. A/C ratios in Figure 14 are between 1.1 and 2 with only one large ratio. The four mud-pellet ratios are beyond 2, but these should not be trusted because of the general behavior of mud pellets. It is also clear that size of spherulites is not an important factor because the A axis varies between less than 10 and 40 without change in ratios. Figure 14 shows little scatter.

Figure 15 shows the A/C ratios for Area 3 in the foothills of the Blue Ridge within the tectonite zone. The spherulite ratios range from 1.18 to 5.3. Layered oöids range as high as, but not as low as, spherulites, and pellets are as high as spherulites but not below 1.90.

Comparison of Figures 14 and 15 shows that Area 3 is geologically far more complex and that the deformation intensities are much higher than in Areas 1 and 2.

Table 4 and Plates 1 and 2 show the average (and range) of A/C ratios for the three oölite belts. A review of 219 oölite samples furnished the data for Table 4. About 30 additional samples were not included because deformation in them is too intense to permit measurements, and oöids can be recognized only partially. Undeformed mud pellets are also excluded. Of the 219 samples, 128 are easily recognized spherulites with sharp boundaries; 56 are spherulites, but with detached outer layers. Thus 184 oölites of the same species can be readily compared.

In Area 1, where deformation is slight, 28 unlayered, sharply defined spherulites were measured, and all other varieties show no deformation.

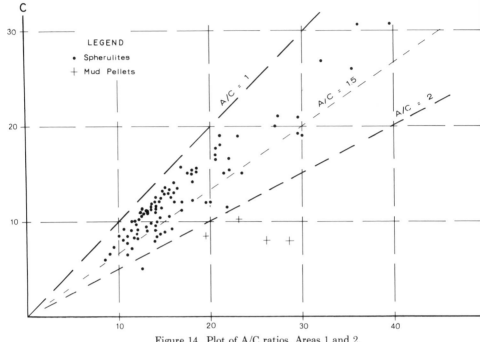

Figure 14. Plot of A/C ratios, Areas 1 and 2.

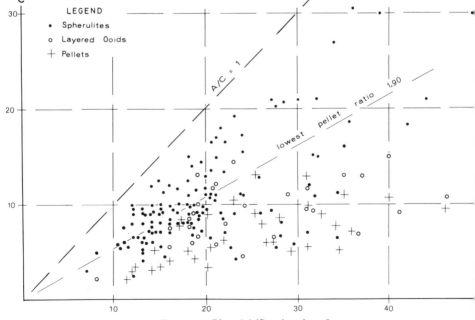

Figure 15. Plot of A/C ratios, Area 3.

Table 4. Areal distribution of oöid A/C ratios, species, and areas.
(Dolomite aggregates are not included. Deformation intensity increases
from west to east.)

West Area 1 West side of Valley	Area 2 West of Massanutten Mountain	East Area 3 between Blue Ridge and Massanutten Mountain
28 spherulites	48 spherulites	52 spherulites
Average 1.16	Average 1.34	Average 2.20
Range 1.11–1.25	Range 1.12–1.68	Range 1.23–5.67
No detached layers	Few detached layers	Detached layers (56 slides) Average 2.57
Mud layers not oriented	Mud layers not oriented	Mud layers (13 slides) Average 2.78 Range 1.42–5.41
Mud pellets not oriented	Mud pellets not oriented	Mud pellets (22 slides) Average 4.48 Range 2.30–6.7

In Area 2, 48 hard-core spherulites were measured; one of these has a detached layer. Mud pellets were either unoriented or only feebly aligned. In Area 3 along the foothills of the Blue Ridge the picture is more complicated. Fifty-two spherulites without layers or mantles were measured and are directly comparable to those in Areas 1 and 2. In addition, in 56 samples of spherulites, layers are detached in varying degrees. Thirteen mud-layered spherulites and 22 slides with pellets were measured.

Spherulites are obviously the most abundant oöids, and also the easiest to recognize under the microscope and in saw cuts. Measurements are comparable and reliable in spherulites either with or without layers. Layered oöids show increases of distance between layers in the direction of maximum elongation, comparable to the increase in thickness between beds in a fold. The remaining oölite samples containing mud pellets or lumps are not comparable in the different areas because they are unoriented and essentially undeformed in Areas 1 and 2.

Within Area 3, the ratios become larger as mud layers and mud pellets are measured. Mud pellets that originally were not spherical would require relatively more deformation before any measurable elongation becomes visible. The wide variety of odd shapes is illustrated in Plates 15 to 18.

Entirely undisturbed are dolomite clusters and chert nodules. If defor-

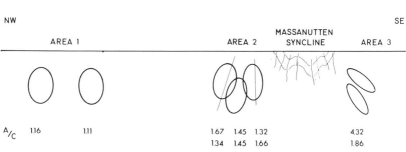

NW   SE

WINCHESTER   BERRYVILLE   SHENANDOAH R.

A/C 1.14   1.19   1.34   2.94   1.96
        1.16   1.21   1.35   2.75   2.00
        1.20   1.25   1.18   1.99

NW   MASSANUTTEN   SE
         MT. JACKSON   MTN.   LURAY

A/C 1.12   1.18   1.35   3.18   6.66
      1.11   1.19   1.40   5.67
      1.16   1.25   2.49   2.42   5.84
      1.17   1.16   3.46   2.37

NW   NORTH   MASSANUTTEN   SE
     MTN.   BROADWAY   MTN.   INGHAM

A/C 1.10   1.13   1.28   1.63   4.66 AVE.
                                    4.11 2.72 4.2 3.85
                                  4.57 5.43 5.26

NW   MASSANUTTEN   SE
     AREA 1   AREA 2   SYNCLINE   AREA 3

A/C 1.16   1.11   1.67 1.45 1.32   4.32
                             1.34 1.45 1.66   1.86

Figure 16. Diagrammatic cross section at Berryville, Va., showing orientation of long oöid axes and ratios. Ellipses are not to scale.

Figure 17. Diagrammatic cross section at Luray, Va., and across Massanutten Mountain. Ratios at Luray are higher than at Berryville (Fig. 16) or Grottoes (Fig. 19). Ellipses not to scale.

Figure 18. Diagrammatic cross section at Ingham, Va.

Figure 19. Diagrammatic cross section at Grottoes, Va. (Massanutten Syncline).

40

mation becomes very intense, the clusters fall apart, and fragments are strewn about or are aligned in linear stringers.

Table 4 and Figures 14 and 15 show slightly different ranges, demonstrating the importance of selecting comparable species.

Variations across strike

Four schematic profiles of ratios were constructed about normal to the trend of the Blue Ridge and the Great Valley: from Winchester to Berryville (Fig. 16); North Mountain through Mt. Jackson to Luray (Fig. 17); from North Mountain to Ingham (Fig. 18); and from North Mountain to Grottoes (Fig. 19). The sections are diagrammatic presentations of ratio localities. Some of the localities were projected into the section from a mile or more to the north and south, and the ellipses show the plunge of the long axes, but are not to scale.

The Winchester–Berryville section cuts across only two oölite belts; all others show three belts. Ratios are lowest in Area 1 (av. 1.16), somewhat higher in Area 2 (av. 1.34), and highest in Area 3, (av. 2.20). A similar increase was described in Maryland (Cloos 1947) where the increase is more rapid and less confused, but along the Potomac River the westernmost oölites are more deformed than are those in the center of the valley.

Perkins (1967, pp. 95–101) describes a section with seven localities, in which the highest ratio is 2.71 east of Massanutten Mountain, and the next highest is 2.64 from the belt west of Massanutten Mountain. These samples agree very well with my observations, but not with the averages of the 219 samples collected in the three belts (Pls. 1, 2). The variations are a puzzle that can be solved only by painstaking sample collecting around entire folds, on overturned limbs of asymmetrical folds, and at structurally well-known localities.

Perkins states (1967, p. 101): "more specimens of known structural locations would be required to determine precisely the variations in deformation, but the limited data at hand indicate some irregularities in the amount of deformation across the Valley." I wholeheartedly agree but believe that the variations across the valley are well established in Plates 1 and 2 and Figures 16–19.

Variations within areas

In Area 1 deformation is near the lower limit of orientation and measurability. Ratios range between 1.11 and 1.29, with only one higher value of 1.46 at Parnassus. The average ratio is 1.16. There are also undeformed oölites with random orientations where elliptical oöids are parallel to bedding. Mud pellets are so little deformed they are not usable.

Area 2 along the west flank of Massanutten Synclinorium consists of two parts: a northern anticline which plunges to the south near Broadway (Pl. 2), and a second one which is nearer Massanutten Syncline between New Market and Staunton. Average ratio for the belt is 1.34. The northern portion shows an average of 1.39 for 16 samples (range 1.13–1.68), and the southern portion shows an average of 1.33 for 15 samples (range 1.15–1.67). This difference is too small to be diagnostic. Many more measurements would be necessary to detect a pattern.

Area 3 shows a variation pattern complicated by the following factors:

1) Greater deformation intensities, as shown in the averages, cause more variation because of differences in the physical properties of all rocks and their components such as spherulites, dolomite, mud, matrix, quartz grains, chert, recrystallization phenomena, and cleavage and lineations.

2) Variations are noticeable along the strike (Appalachian trend).

3) Variations occur between the normal, upper, and overturned lower limbs of asymmetrically overturned folds.

4) The Blue Ridge foothills are less well exposed than other portions of the valley.

5) Faults, thrusts, and shear zones complicate that area, cut it into fragments, and tend to obliterate oölites.

Between the Potomac River and Lexington, Virginia, Area 3 may be divided into three parts: a northern section between the Potomac River and Front Royal, a central portion from Front Royal to south of Elkton, or the south end of Massanutten Mountain, and a southern portion from there to Lexington.

The averages and ranges are as follows:

Section	Average	Range
North	2.53	1.23–5.40
Center	3.42	1.42–5.67
South	2.20	1.20–5.54

The range of ratios is almost the same in the three sections, but the average is far higher in the central portion. The Massanutten Synclinorium plunges south at its north end near Strasburg and to the north at its south end between Harrisonburg and Elkton. Its deepest portion is therefore like a boat and preserved only in the center. Its northward and southward continuations are eroded. Possibly that portion of the Great Valley where the Silurian is still present has been more intensely compressed than to the north and south.

Massanutten Syncline and the belt of Martinsburg Formation narrows to the south and north. It is only a little more than 2 miles wide where the Potomac River is crossed by Martinsburg Shale. The synclinorium is a tight structure which continues downward into a tight upright syncline of

Martinsburg Shale and widens upward to include the Silurian and a portion of the Devonian.

Differences between limbs of folds in Maryland have been described (Cloos 1947, p. 880), and they are as distinct and even more noticeable in the Blue Ridge foothills south of the Potomac River.

In Area 3 three averages were calculated for A/C ratios in the upper limbs (2.23), vertical limbs (2.41), and overturned limbs (3.84). The difference between the first two is slight, but in the overturned limbs ratios are much higher. This situation is illustrated in Figure 28, which is a construction similar to that in my earlier paper (1967, Fig. 8).

Zoning within the east belt is suggested in the Luray area; in localities south of the city, west of U.S. 340, and to the east on the road to Antioch church, ratios increase eastward. This may be due to the position of oölites in folds, to an increase of strain, or to zones of closely spaced cleavage in limestone. Such zones have been described above under "Extremely deformed oölites." The zones parallel the general trend, are about 50 to 100 feet wide, and contain highly schistose limestone which is also intensely lineated. In thin section the limestone is streaky, and only rarely can a suggestion of a remnant spherulite be detected. Plates 35 and 36 show schistose limestones, but oölite remnants are still visible. On either side of these zones, the limestone and oölites are less schistose and normal. An excellent outcrop of an entire zone is along the Baltimore and Ohio Railroad above Harpers Ferry in West Virginia, where the tracks cut across the Tomstown Formation. Bedding is vertical and indicated by thick dolomite layers. Cleavage dips 15°–30° E, and the lineation is in the cleavage in direction 120. Other good examples are north of Front Royal on U.S. 340, 1 mile east of Double Tollgate, and west of Elkton on U.S. 340 at the west end of the Shenandoah River bridge.

In the railroad cut at Ingham, three easily recognizable but strongly deformed oölite beds grade westward into a highly schistose zone. Within the zone the limestone falls apart in slate-like plates on cleavage, not bedding. West of the zone, along U.S. 340, the limestones are massive, thickly bedded, and without cleavage. Whether or not faults follow the zones is not clear, but formations are not interrupted. Neither the zones nor faults are shown on the geological maps, nor are the zones mentioned by King and others.

VIII

Other Strain Markers

Quartz pebbles and fossils have not been used in this study because oölites are more satisfactory, more abundant, and occur at strategic locations. Worm tubes are common in the Antietam Formation, very abundant at a few localities, but otherwise unevenly distributed and therefore less useful. Chlorite blebs, amygdules, and pebbles occur in the Catoctin Volcanics. Amygdules are abundant in flow tops and flow-top breccias and are like guide fossils for flow tops. They show primary elongations and arcuate arrangements, but are only slightly distorted. Their distribution is also too limited. Elongated and well-aligned chlorite blebs occur in thousands of outcrops (Pls. 1, 2, 67, 68). They are excellent as direction markers, but questionable as strain markers because one axis is very long, one is intermediate, and the third is rarely measurable because the blebs are paper thin. Their original shape is unknown, and they may indicate slickensides in cleavage surfaces.

Chlorite blebs in greenstone are shown on Plate 67. Similar blebs in tuffaceous beds are shown on Plates 68 and 69.

Ratios can be measured only in *ab* surfaces in which blebs are found as AB sections. The C dimensions are so thin that an A/C ratio may be over 100. This can hardly represent the strain but must include slippage on *ab* surfaces.

For comparison some ratios were measured.

Dimensions of blebs in volcanics:

A	B	C	A/C ratios
21	3.3	.1–.5	175 ± 25 Greenstone
16	5.0	.1–.5	160 ± 32 "Pumice"
8	3.5	.1	±80

The values show the typical variability. Only about 100 measurements illustrate the excessive A/C ratios. Elongated black spindles in the gneiss (Pl. 63) also suggest strain markers, but are also limited as markers. The original shapes may represent isometric garnet crystals that were chloritized. The elliptical black streaks also contain crushed quartz. Their ends are frayed and vague and in thin sections they are seen only as discontinuous micaceous layers whose ends are indefinite.

An unsuccessful attempt was also made to measure epidosite lenses, but they are too irregular, rarely elliptical or lenticular, and are from a few inches to dozens of feet long. The large ones are undeformed, massive centers which deflect the cleavage and cause it to bulge.

Inasmuch as oölites are so well distributed, abundant, and of relatively uniform size and substance I concentrated on them and used other possible strain markers only as indicators of directions.

IX

Lineations

Of the 15 lineations discussed in 1946 (E. Cloos 1946) most are not used in this paper; others, not included there, such as striae on slickensides are now being used. This does not mean others do not exist or may not be important, but not all lineations are of equal interest in this study.

Lineations have been discussed recently by several authors. Ramsay discusses it, Whitten (1966, pp. 264–321) devotes a chapter to "linear structures," and Fisher *et al.* (1970, p. 453) list lineation in the index on 54 pages.

Whatever the interpretations have been is of little interest here because the phenomena described below can be seen in the field and have been measured in thousands of outcrops. Their geometric relations are clear, very consistent, and they came into being by a mechanism which is open to interpretation, but the facts are astonishingly simple.

A structure that can be traced from the northern termination of the Blue Ridge in southern Pennsylvania for 400 miles to the Great Smokies in Tennessee with surprising consistency deserves some attention in local reports and maps, especially because lineation and its orientation in relation to tectonic transports has been the subject of controversy for decades.

Some rocks are very obviously lineated, others less so. Some of the measurements deviate from the maximum and prove only that when 40,000 to 50,000 cubic miles of rock are moved not all particles move in parallel paths or at the same rate. If metamorphic processes are added and temperature differences and time taken into account, it is most surprising that maps such as Plates 1 and 2 show so many consistent elements in such a large area over such a variety of rocks with widely differing properties.

I have used the lineations listed in Table 5.

Lineation in gneiss and volcanics

This lineation has been mentioned frequently, and its direction is very consistent over wide areas (Cloos and Hietanen 1941, p. 83, Pls. 8, 9; E.

Table 5. List of lineations used

A. Lineation in (Precambrian) gneiss and volcanics: mineral patches, blebs, clusters, aggregates, elongate spots or single parallel elongate mineral grains.

B. Lineation in overlying sediments: elongate quartz grains, streaks, elongate spots, and patches.

C. Long axes of deformed oöids grading into streaks in intensely deformed oölites.

(A, B, and C are penetrative, not limited to surfaces, but within the entire sample and not always easily distinguished from striae on slickensided surfaces.)

D. Striae on surfaces, slickensides.

E. Parallel mineral growths of actinolite and prismatic minerals on fractures.

F. Growth of quartz, calcite, chlorite, feldspar, actinolite across fractures. These are called "rods" for convenience.

(A to F are statistically in the ac plane normal to fold axes and general trend as seen on all geologic maps of the area. That trend is here called "b" = B for convenience.)

G. Crenulations at a large angle to striae on slickensides, or across long bleb axes in volcanics or other lineations included under A to F.

H. Intersections of S-surfaces such as bedding and cleavages, several cleavages, or joints, cleavages, and bedding.

I. Fold axes.

(F to I are very valuable and far less consistent. It might be sensible to determine the ac plane statistically by several hundred measurements and to construct b normal to ac. This b may deviate from fold axes at points but not over large areas and many readings.)

Cloos 1946, 1947, 1950, 1951, 1964; Reed 1955; Nickelsen 1956; Balk 1952; Kvale 1966; and others). It has been subject to much needless controversy (Turner 1957).

Lineation in gneiss is very common and widespread. An example is shown on Plates 63 to 66. Almost all outcrops of gneiss along the Skyline Drive show parallel orientation of dark patches or mineral alignment. In some places it is the only good structure. For instance, on and around Old Rag Mountain, Shenandoah National Park, east of Skyland, two assistants and I spent an entire day on the large bare rock surfaces and were unable to measure anything but one type of lineation. The Old Rag Granite is a lineated rock to the exclusion of almost any other structures; even dikes and joints are scarce. (See also, Reed 1955.)

In the Catoctin Volcanics a similar lineation is very common and consists of chlorite or sericite blebs that are narrow and long within cleavage planes and very thin normal to cleavage (Pls. 67–70). The long axes also dip east or southeast. In Plates 1 and 2, the orientations of blebs are shown as long narrow black ellipses with dip of long axes along the Blue Ridge. Each map entry is an average from 50 to 100 readings in the field.

Few outcrops in Catoctin Volcanics are without lineation, but it varies greatly in intensity and spacing in an outcrop or sample. Massive greenstone shows no blebs. Completely free are the bodies of epidotized greenstone, which are also free of cleavage, but are extensively fractured or mylonitized.

Flow-top breccias, massive flows, flows showing cooling columns, and generally nonschistose portions of Catoctin Greenstone are free of elongated blebs or may show very small and barely elongated ones. Where massive lenticular greenstone blocks are surrounded by intensely schistose layers of greenstone the interior of the lenses may be nonlineated and not schistose.

The distribution of lineation is thus rather irregular and depends on what happened to the volcanics. Massive greenstone is as common as schistose greenstone, but the distribution has not been systematically mapped.

Lineation in sediments

Lineation in sediments may be an alignment of small blebs of sericite, weathered feldspars, shale pebbles, quartz grains, or dark spots, or it may be streaking and mineral alignment in phyllites, or elliptical quartz grains in quartzites. In the limestones, lineation is seen as patchy long streaks of varying composition or grain sizes, or color. By no means do all sediments show penetrative lineation of components, and very massive quartzites, like the Antietam, are practically free of lineation.

The Loudoun is well lineated at many places (Pls. 72 and 73).

In the Harpers Formation, lineation is also common and normal to the fold axes and the intersection of cleavage and bedding. It consists of "rust spots" or weathered pyrite, magnetite clusters, streaks of mica, and particle clusters. It is not easy to find.

In the limestones, lineation reaches as high as the Athens above the Beekmantown, but it is not common. It seems limited to schistose zones which are up to 100 feet thick. Within such zones the limestones are calc schists with closely spaced and intensely lineated cleavage planes. Schistosity and lineation cut across bedding. Very prominent, well-exposed zones occur west of Elkton, at the west end of the Shenandoah River bridge, in the Ingham section along the railroad, and at the intersection of U.S. 11, opposite Howard Johnson's restaurant.

Long oöid axes

These are shown on Plates 1 and 2 as ellipses of three kinds with dotted centers and arrows and figures for dip. Most of the photomicrographs also show the long axes and are AC equal *ac* cuts.

The arrangement of long oöid axes is the most prominent and consistent penetrative linear direction which can be measured in the limestones. It has been discussed under oölites above and is again elaborated on below in the discussion of oölites and deformation. It also has been reviewed by Whitten (1966, pp. 277–81) and Ramsay (1967).

Striae on surfaces, slickensides

Striated surfaces have been noted by miners and geologists for well over a century. Textbooks mention slickensides (for instance, Hills 1963, pp. 175–77; Billings 1954, p. 149), but they are not generally found useful, except that they are mentioned where bedding planes show striations due to flexural slip, or faults are covered with striae that indicate displacement directions. Systematic surveys of striations are rare mainly because the directions of striae seem random, and conflicting directions occur on closely spaced layers. A. Kvale (1966) reported on striations on slickensides of the Gotthard Massif which resembles the Blue Ridge.

During the field work along the Skyline Drive the coincidence of the orientations of lineation in the Catoctin Volcanics and the Chilhowee Formations and of striations on slickensided surfaces became rather striking. Lineation is very consistent, and striae on slickensides are essentially in the same vertical plane in which the lineation occurs (ac). After this became fairly well established a systematic study of striations was undertaken, and the orientations of 3,238 slickensided surfaces and of 3,710 striae were measured in the field. All formations between the gneiss and the Devonian are represented, but slickensides are most common in the Catoctin Volcanics of the Blue Ridge. Table 6 shows the distribution of measurements in the stratigraphic column.

The distribution of observations is influenced by several factors: exposures, accessibility, occurrence in different formations and rock types, time spent in the areas, and preservation of slickensides and striae.

In limestones, slickensides are quite perishable. Farming in the valleys tends to destroy exposures, whereas the Blue Ridge rocks are protected in the National Park and are also far more resistant. Also most favorable for detailed observations are artificial outcrops which are most abundant along the Skyline Drive and Blue Ridge Parkway and access routes. Finally, hillwash from the Blue Ridge covers the limestones in broad terraces along the Shenandoah River and the slopes of the Blue Ridge.

The observations, therefore, do not represent a typical natural pattern but largely depend on circumstance. In all areas additional data can be

Table 6. Field measurements of slickensides and striae

	Slickensides	Striae
Devonian	55	56
Tuscarora	140	156
Martinsburg	189	225
Limestones	251	245
Chilhowee	520	653
Catoctin Volcanics	1,632	1,901
Gneisses	451	474

gathered in almost any desired quantities. I feel, however, that the data are sufficient to show the consistency of the observed pattern.

Striae are caused by slippage on surfaces such as faults or bedding planes, and they are limited to the surfaces on which they occur; a lineation within the rock is due to penetrative deformation and recrystallization. Chlorite blebs in volcanics, for instance, are cut by fractures with striae that are at angles to the rock lineations; in this case the lineation is pre-slickensides, but there is no indication of the time interval.

Striation on cleavage

Many cleavage surfaces are lineated, and it is not easy to distinguish striation on surfaces from lineation in the adjacent rock. They grade into each other, and their orientation is the same. The two structures are so common that early in the investigation I may have measured striation in cleavage. Later on a separation was made between blebs in the rock and striated surfaces.

Where cleavage bends around massive greenstone or epidosite lenses the surfaces are striated, there is mineral growth, and the appearance is that of slickensides.

Plate 76 shows curved cleavage in greenstones. The process that produced bleb orientation may well have continued into formation of cleavage, and after cleavage existed it served as a dislocation surface on which blocks moved relatively to one another, resulting in slickensides.

Striae occur on the following kinds of surfaces: (1) cleavage or parallel cleavage in volcanics and gneiss; (2) bedding in sediments; (3) closely spaced curved surfaces, probably gradational with cleavage; and (4) small cross-cutting fractures, mineral-coated large cross-cutting joints, some faults.

Striated bedding

Bedding surfaces in sediments are commonly striated in either one or several layers. This striation occurs in all formations, including the Devonian, and orientation is very consistent. Most of the slickensides are covered with secondary quartz or calcite which may make veins up to 1 inch thick and are composed of several layers with different orientations of striae. This suggests several generations of movements in different directions, accompanied by mineralization. Plate 77 shows a slickensided bedding plane in the Weverton Formation. Not all bedding planes are striated where they are closely spaced, but slickensided bedding planes usually are not more than 3 or 4 feet apart. In thickly bedded Tuscarora Sandstone almost every bedding plane is striated, but the surfaces are more weathered than in the Martinsburg Shale, for instance. A systematic study of spacing of slickensides, lithologies, and bedding thicknesses was not made, but would be well worth while.

Cross-cutting fractures

Cross-cutting fractures are abundant but are not commonly slickensided. They were not systematically studied but were measured where slickensides occurred (Pl. 80).

Fibrous mineral growth in fractures

Many slickensided surfaces are scratched surfaces, and others are accompanied also by mineral growths in fractures; but the patchy, streaky, or grooved orientation is related not only to growth of fibrous minerals but also to accumulation of minerals in groups and patches and to movement on the surfaces.

Minerals grow on surfaces and look like slickensides. They need not be due to displacements but are due to growth in fractures. This is particularly common in the epidosite blocks and lenses of the Catoctin Formation. Many fractures are as small as those shown on Plate 79.

Figure 20 shows a sample of epidosite with radiating fractures. Two kinds of growth occur: actinolite and quartz parallel surfaces and fibrous

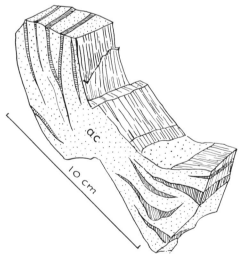

Figure 20. Drawing of sample of fractured epidosite (compare with Pl. 79). An ellipsoidal, brittle, epidosite lens has been rotated (rolled) in greenstone. Tension fractures opened and grew wide near the periphery, center not fractured. Mineral growth in fractures is quartz, actinolite, and epidote, either normal or parallel to fracture, but always parallel with front surface ac.

quartz, actinolite, and chlorite across the fractures. The small sample is like the regional orientation of fibers, which is as rigid as in the small sample of Figure 20.

Many fractures show growth of fibrous quartz and actinolite across as well as within the fractures (Pl. 79). The two growth directions occur together. The phenomena cannot be separated, but this growth orientation is not slickensiding.

Where quartz-chlorite-actinolite rods protrude from a rock surface the orientation looks like a small portion of a slickensided surface. This illusion can be verified by a study of sections across such fractures (Pls. 81 and 82).

Growth of quartz rods is so common in the Catoctin Volcanics that almost all outcrops show it in large and small fractures. Fibrous mineral growth is not limited to the volcanics but is widespread in the gneiss and the Paleozoic rocks. Frequency depends on the occurrence of joints in the formations. The Catoctin Greenstone is highly jointed because it was folded and fractured. Within the greenstone the epidosite nodules and lenses are the most fractured portions and also show the most abundant mineral growth.

It may become difficult to distinguish growths from fracturing of larger vein fillings. For instance, Plate 83 shows quartz fibers filling an irregular opening. Extinction suggests about 10 or 12 large quartz crystals that have been shredded. The fibers are parallel to innumerable fine actinolite needles within the larger quartz grains. The laths have then been bent and "folded" on two fractures and also along their contacts. Some laths are granulated, and many hundreds of small grains are elongated and at an angle of about 45° to the long axes of the laths. Very small actinolite needles also grow in this direction.

The sequence in Plate 83 may have been as follows: A fracture was filled with large quartz crystals; these were then broken into rods or laths; actinolite needles grew parallel to laths; the quartz laths and needles were bent into flexures or were broken with some recrystallization in the fractures; quartz became sheeted across its crystallographic axis more or less parallel with the fibers. Many of the laths are undulatory, and some large-grain laths are granulated and individual grains are elongated in a new direction about 45° to the laths. This history indicates that the needles and laths are not growth on the wall of a fracture but have become fibers due to deformation after vein filling.

The sequence of quartz growth, rodding, and formation of fibers can be seen in flow-top amygdules filled with quartz. Plate 84 shows elliptical amygdules from the flow-top breccia at Big Meadows, Shenandoah National Park. Quartz is not even undulatory, in spite of the ellipticity of the amygdule, because the distortion of the amygdule is primary and due to flow. A few large quartz grains filled the voids.

The impact of deformation is illustrated by Plate 85. An amygdule is filled by several large quartz grains which are strained uniformly across the amygdule and not in relation to quartz extinction as is common in undulatory quartz. Plate 86 shows needles that radiate from a point. The amygdules show that the vein filling was by large unoriented quartz grains that later became needles. A sample of intensely deformed vein quartz is shown in Plate 87.

Crenulations

Crenulations on slickensides are at a large angle, but not necessarily normal to the striae. They affect the striae which are bent by the small folds and must therefore have existed prior to the formation of the small foldlets. Wave lengths are of the order of centimeters, and amplitudes are also small. Slickensided surfaces are wavy, and the folds are limited to the mineral coating of the slickenside. The adjacent rock is not affected. It is, therefore, quite different from the fracture cleavage or other superimposed cleavages which are functions of shear planes that cut across the rock and several slickensides or bedding planes.

Crenulated slickensides must be related to the movements on the surface, if only the immediate thickness of the slickenside and mineralized surface is crenulated, because the curved and slickensided surface wraps around more massive portions of greenstone, epidosite lenses, or dolomite boudins.

Intersections of surfaces

Intersection of cleavage and bedding is most frequently observed and described, but it cannot be observed in the gneiss or Catoctin Volcanics except in interbeds between flows. In the Chilhowee Formations, the limestones, the Martinsburg Formation, and stratigraphically higher strata the intersection is most commonly parallel to fold axes, but not necessarily so. Wickham (1969) has described the intersection in the Front Royal area, and similar conditions prevail elsewhere.

Intersection may be either bedding and axial-plane cleavage, or bedding and/or axial-plane cleavage and a second cleavage such as fracture cleavage or microcleavage (Stevens 1959) or other superimposed cleavages. All are well known and common in such rocks as the Harpers Phyllite and other incompetent formations, but they also occur in limestone interbeds. Intersections were not as completely pursued as oölites and lineations, but 200 of a total of 400 measurements are shown in Figure 23, including all formations between the Precambrian crystallines and the Devonian in the area south of the Potomac River to Lexington.

Deviations are rare and may be due to faulting or distortions of folds. A pattern of superposition of different directions cannot be recognized.

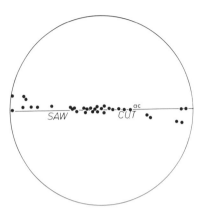

Figure 21. Diagram of 20 fiber orientations of sample shown in Figure 20. Fibers are solid dots. A girdle almost parallel with saw cut. Slight deviation due to cutting sample. Fibers are like spokes of a wheel in *ac*.

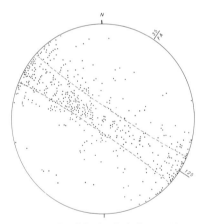

Figure 22. 530 bedding-plane poles. Maximal girdle at 125 normal to fold axes and intersections of Figure 23. Scatter is due to plunging folds which are particularly well shown on Elkton quadrangle map (King 1950).

Fold axes

Fold axes were directly measured where hinges are exposed. They range in size from small folds in phyllites to larger folds in dolomite beds in limestone or the Chilhowee Formations. All are folds of bedding, not of cleavage.

Only about 50 hinges were measured excluding cleavage-bedding intersections. Folds are common, but well-exposed hinges are not seen as often as cleavage–bedding intersections. Their orientation is identical to that of the intersections in Figure 23.

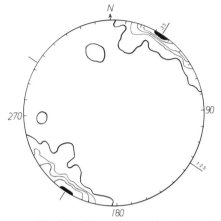

Figure 23. 200 cleavage and bedding intersections. About 100 points are direct measurements; the others were constructed from cleavage and bedding in outcrop. Measurements
were grouped in lots of 40 as they occurred in chronological order in notebooks of the last
10 years. Each group was first plotted separately. When the consistency became clear all
points were assembled into this figure.

Origin of "a" lineation

Long axes of the distorted oöids and especially of mud pellets are
directions of extension, growth, and flow, generally perpendicular to
fold axes. The chlorite patches in greenstones and black clusters in the
basement gneiss are also normal to fold axes and in the same ac planes
as the long oölite axes. Growth of quartz rods and actinolite needles are
in that ac plane, and striae on slickensided fractures are also in the ac
plane and normal to b.

This general agreement of structures suggests that lineation ranges between mineral-growth orientation, extension, flowage, and slickensiding
and that the lineation a shows itself in different forms depending on the
host rock, the intensity of deformation and displacements, the available
material, and the amount of recrystallization that can take place at a
given locality.

The a lineation is a prevailing direction but in different forms. Considering the large displacements that must have occurred as the basement surface was turned from horizontal to vertical and considering the
folding that accompanied the upturning, a is, in the broadest sense, the
direction of tectonic transport, slippage, growth, or extension, and probably all of these. It has been effective at different times and in different
forms with a constant orientation within a plane that is normal to the fold
axes. The controversy about lineation in a or b arose largely because of

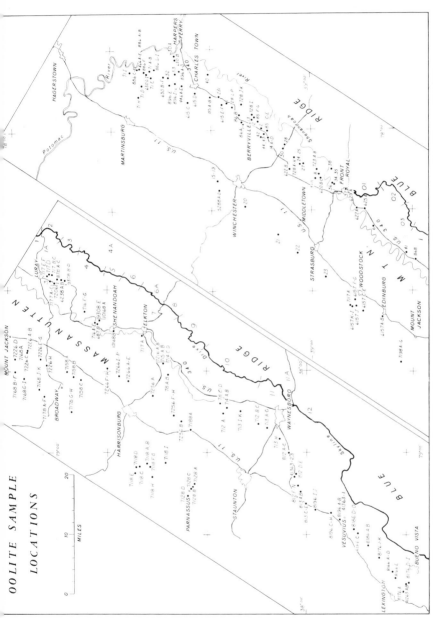

Figure 24. Oölite sample localities. Subareas 01–12 along Skyline Drive correspond to Figures 31 to 46.

lack of knowledge and communication between geologists who have studied different regions.

The situation along the Blue Ridge from Pennsylvania to Roanoke is so easily seen in the field and so consistent that there is no need to add to that discussion. Similar lineations have been described by Kvale (1966), Balk (1952), Fisher *et al.* (1970), and others.

X

Cleavage

The Precambrian gneisses are schistose, especially in shear zones (Pls. 63–66), but massive gneiss abounds. The Catoctin Volcanics are rarely without cleavage, but massive greenstone is common; the Chilhowee Sandstones (Weverton, Antietam) show cleavage, and the Harpers Formation is so schistose that bedding is hard to see at many places and is only a remnant discoloration or grain-size change and has ceased to be mechanically significant. In the limestones cleavage is very common in zones and also where deformation of oölites results in an A/C ratio of 2 or more. Dolomites and many younger limestones like the Beekmantown show no cleavage. The Athens, however, is intensely cleaved, and the Martinsburg Formation is a slate in many portions. Above the Martinsburg cleavage is common in all incompetent shaly formations like the Wills Creek (Silurian) or Romney (Devonian) and up to the Carboniferous shales.

Cleavage (S_1) was bemoaned early by Rogers (1884, p. 158), who notes that the dip of bedding is difficult to ascertain because of the cleavage. Cleavage has been mentioned, measured, illustrated, and studied by a number of authors, for instance Fellows (1943), who illustrated cleavage in the Weverton. Cloos (1951, p. 144) described it in the report on Washington County, Maryland. Reed (1955, 1969) described cleavage from the Luray area in Catoctin rocks. Nickelsen (1956) measured cleavage south of the Potomac River in Virginia. More recently cleavages are discussed by Lee (1961), Perkins (1967), Dean (1966), and Wickham (1969), who studied it in some detail. Fisher *et al.* (1970) show 46 cleavage entries in their index, in addition to foliation and schistosity.

It seems agreed that cleavage is prominent in the Blue Ridge and in the Great Valley to the west, that it dips east generally (Rogers 1884, to Wickham 1969), that it is very useful in determining top and bottom in overturned limbs of folds, and that it becomes less distinct proceeding across strike from southeast to northwest. It is also recognized that its orientation becomes less rigorous and uniform as axial cleavage fans (Dean 1966, p. 85; Cloos 1951). Fellows (1943, p. 1430) drew a boundary between tectonites and nontectonites, separating the Blue Ridge Rocks from those without cleavage and lineation to the west, and called it the "tectonite frontier."

In the oölites cleavage is related to deformation intensities, and it seems reasonable to assume that this is also true in the noncarbonates.

In the Catoctin Volcanics cleavage does not occur as simple planar structure, but it flows around more massive portions on a large scale. Such "bulging cleavage" must have been formed in a plastic stage because cleavage is accompanied by mineral orientation. Slickenside striation and orientation of curved surfaces fit snugly as separate curved "shells" and are not fractured as they would have been if the shells were fitted in a brittle state. Where fractures occur they are within the lens around which the cleavage is wrapped. The fractures can at times be related to the movement as shown in Figure 20, where displacements show the direction of rotation. The top has moved westward over the bottom portion. The fractures are filled with actinolite, quartz, and epidote. Bulging cleavage is common only in the Catoctin Volcanics, where massive greenstone, epidote lenses or "gobs," and flow-top breccias are numerous. It is less common in the limestones where dolomite beds are fractured, boudinage is common, and blocks have been pulled apart and cleavage or bedding curves around brittle blocks (Wickham 1969).

In oölites cleavage becomes visible in thin sections as a parallel arrangement of minerals or components, or as surfaces that transect thin sections. Mere parallel arrangement of elliptical oöids is not here called cleavage. Oölites are very useful in relating strain and cleavage phenomena. The most sensitive indicators of cleavage are mud pellets, muddy portions of sections, and mud layers around oöids. These are also the finest-grained portions of the oölite. Almost immune to cleavage are the crystalline matrix of oölites, dolomite clusters or grains, and chert.

In Table 7 I have listed the photomicrographs of thin sections shown in Plates 3 to 62 in order of increasing A/C ratios. Descriptions are by key properties such as spherulites, mud, mantles, centers, pebbles, etc. Plate and figure numbers are shown, and sample numbers are listed for identification.

Cleavage is as selective in oöids as in a sequence of alternating competent and incompetent sediments. It is first seen as parallel arrangement of calcite crystals in mud pellets, muddy matrix, or mud centers of spherulites at A/C ratios above 1.40. It remains feeble and can be seen only at high magnification up to ratio 1.56. At ratios above 2.00 it is distinct and at 2.5 it cuts across oöids; above 3.00 the oölite is megascopically cleaved, and all components but the matrix participate. Sparry calcite shows calcite cleavage, but rarely abundantly except at the ends of compressed oöids and in growths beyond the ends of oöids.

In following the photomicrographs (Pls. 3–62) from low to high ratios it is obvious that cleavage is a function of deformation intensity. Table 7 illustrates the interdependence of cleavage and ratios and the fact that cleavage in oölites is prominent only in Area 3 adjacent to the Blue Ridge.

Table 7. Strain and cleavage in oölites.

Ratio A/C	Species	Area	Plate and Sample No.
1.14	Spherulites	2	Pls. 56 and 57, 5288B
1.14	Spherulites	2	Pls. 45 and 46, 5288D
1.16	Spherulites, closely packed, layered	2	Pl. 9, 7256Ea
1.18	Spherulites, layered	2	Pls. 3 and 4 7256Cb
1.22	Spherulites and dolomite grit	3	Pl. 42, 8176Ea
1.25	Spherulite, single	3	Pl. 47, 630D
1.31	Spherulites, radial fractures		Pl. 49, 420
1.35	Spherulites, dolomite pebbles	2	Pl. 21, 7226B
1.38	Pellets with growth; quartz	2	Pl. 20, 7158Cb
1.40	Mud pellets with quartz	2	Pl. 27, 7226Ab
1.41	Spherulites, pebbles. Cleavage in mud	2	Pl. 29, 7266Ab
1.44	Spherulites, fracture pattern	2	Pl, 7, 812Hb
1.48	Spherulites, mud pellets, fragments	2	Pl. 31, 7266G
1.52	Spherulites, pellets, pebbles	2	Pl. 19, 7266Ma
1.53	Mud centers in spherulites	2	Pl. 18, 7226F
1.53	Mud centers in spherulites, coalescing fractures. Cleavage faint	3	Pl. 37, 8176Ga
1.56	Spherulites, shells, growth, fracture pattern	2	Pl. 11, 7256Fa
1.68	Spherulites, on mud; pebbles. Cleavage in mud	2	Pl. 8, 7226G
1.71	Spherulites, stylolites; mud	3	Pl. 33, 713F
1.80	Spherulites, mantles, pellets. Cleavage in pellets	3	Pl. 12, 466A
2.00	*Spherulites, mud. Cleavage in mud	3	Pl. 54, 344
2.00	Chert; matrix mud. Cleavage good	3	Pl. 41, 58–14B
2.04	Spherulites, fracture pattern. No cleavage	2	Pl. 58, 58–16
2.07	Spherulites, bedding, radial structures	3	Pl. 6, 85G2b
2.09	Spherulites, mantles, pebbles. Cleavage = needles	3	Pl. 14, 886Cb
2.09	*Spherulites, mantles, dolomite, grit, mud. Cleavage	3	Pl. 34, 886Cb
2.14	Dolomite, mud mantle. Cleavage	3	Pl. 22, 8166Ea
2.20	Spherulites, pebbles, mud. Cleavage in mud	3	Pl. 39, 813A4
2.37	*Spherulites. Cleavage in matrix		Pl. 50, 371A
2.40	*Spherulites, mantles. Cleavage in mantles and matrix	3	Pl. 13, 713Gb
2.71	*Pebbles, pellets. Cleavage in pellets and centers	3	Pl. 30, 813A2
2.91	Spherulites, pebbles. Cleavage in spherulites and mud	3	Pl. 38, 886G2
3.00	Spherulites, mud. Cleavage in mud only	3	Pl. 17, 466Ea
3.12	Spherulites, fragments, mud. Cleavage in mud	3	Pl. 32, 620C

*Indicates particularly good cleavage in photomicrograph.

Table 7 (*continued*)

Ratio A/C	Species	Area	Plate and Sample No.
3.38	* Spherulites, pellets. Cleavage in centers and pellets and matrix	3	Pl. 51, 344
4.57	* Spherulites, mud, dolomite. Cleavage in spherulites, mud, matrix	3	Pl. 23, 687Ba
4.70	* Quartz layers. Cleavage strong across slide	3	Pl. 61, 8166A
4.72	* Streaky limestone, spherulites. Cleavage across slide	3	Pl. 36, 896B
5.54	Spherulites, mud matrix. All with cleavage	3	Pl. 24, 8166Ba
5.67	* Spherulites split. Cleavage strong	3	Pl. 62, 717C
	* Specimen spherulites with slickensiding. Cleavage strong	3	Pl. 62, 717C
	Streaky limestone. No spherulites. Growth on quartz. Cleavage	3	Pl. 60, 58–6

* Indicates particularly good cleavage in photomicrograph.

A second cleavage without mineral orientations due to growth cuts the flow cleavage, makes chevron folds, crenulations on S_1, and displaces it. It has been called fracture cleavage (Cloos and Hershey 1936) and slip cleavage and is rather common in the Harpers Formation at Harpers Ferry (Nickelsen 1956; Dean 1966). Wickham (1969) described it in detail, and Stevens (1959) studied it in a larger area, including the Blue Ridge. Terminology is somewhat cumbersome because the second cleavage is in some situations the only one; elsewhere several cleavages occur. It has been noted in most Appalachian formations from the Pre-Cambrian to the Carboniferous. Like flow cleavage, it is more prevalent near or in the Blue Ridge than to the west.

Lineation and cleavage

In the oölites the time sequence is as follows: Feeble distortion of oöids to triaxial ellipsoids with longest axis A normal to fold axes $b = B$. C is normal to $AB = ab$ plane. Where the distortion increases—mainly approaching the Blue Ridge—the axial ratios become more obvious and at $A/C = 2$ cleavage appears as AB plane equal ab. When ratios increase the limestone becomes a laminated, streaky carbonate schist where lamination is the AB plane of the ellipsoids and where a lineation is obvious in that surface normal to $b = B$. This cleavage and lineation are parallel to those of the volcanics or the gneiss and the sedimentary formations. Since cleavage fans it cannot be parallel across distances. However, when plotted into a net the lineation points are in the same zone as the cleavage poles.

In the flow-cleavage planes lineations listed in Table 5 are particularly common in the gneiss shear zones and the schistose Catoctin Volcanics, and the striae of slickensides are statistically parallel with the blebs in cleavage planes.

There is a gradation from oöid AB to cleavage, to the orientation of long oöids axes in the AB plane, to elongate blebs (A), and to striae on slickensided cleavage planes. The a lineation of the volcanics or the gneiss and some of the overlying sediments, especially in schistose zones, grades into slickensides and may well represent the initial orientation of slickensides prior to the coalescing of orientations into cleavage surfaces.

XI

Orientation and Patterns of Structures

Orientation of oöids

Random orientation of ellipsoidal oöids occurs only in the western belt (Area 1) and here only where axial ratios A/C are less than 1.15. Northwest of Winchester and in the vicinity of Parnassus, orientation of barely elliptical oöids is difficult to measure. Primary ellipticity, even if it is very slight, may parallel bedding, but is difficult to determine. Where axial ratios are over 1.2 a uniform orientation across the thin section is recognizable.

If the 235 available orientations for the longest oölite axes are plotted into an equal-area net diagram, the distribution of points shows a zone 20° wide across the diagram at azimuth 130 (Fig. 25a). Figure 25b is a graphical presentation of the points of Figure 25a. The maximum number is at 130°.

A collective diagram which includes all points seems too general and eliminates all details; thus three additional diagrams were made for the three oölite belts (Fig. 25c, d, e).

In addition all AC sections were plotted into small orientation profiles with field data for bedding, lineation, or cleavage. Some of these diagrams were used in constructing Figures 26–28. Their mutual positions were combined into schematic sections.

Oöid orientations in folds

Most unfortunately I have not found a complete fold in which it was possible to measure oölite distortion in all parts of the fold and in several beds within one fold. All data here presented are from fold fragments and distributed as follows: normal upper limbs were found in 23 stations in Area 1, 32 in Area 2, and 54 in Area 3; one fold crest was found in Area 1, and 11 in Area 3; overturned limbs were found at 76 stations in Area 3, but not elsewhere. Reconstructions of folds from fragments meet with considerable difficulties and some assumptions, but some simple facts emerge.

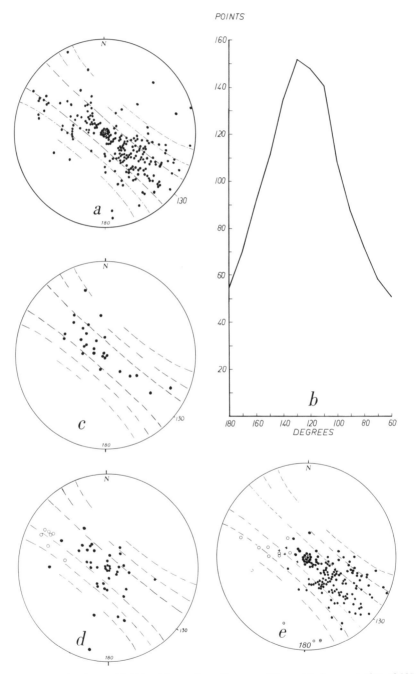

Figure 25. Oöid axes (a) 235 oöid axes for entire area; (b) graphical presentation of 235 axes in zones from 180° to 60° azimuth. Maximum at 130°; (c) axes in Area 1; (d) axes in Area 2; (e) axes in Area 3.

Table 8. Key to orientation diagrams

Figure	Structure elements	Number of observations
21	Epidosite fractures in sample 8227 B	20
22	Poles of bedding	530
23	Cleavage and bedding intersections	200
25 a	Oöid axes	235
b	Graphical plot of a	
c, d, e	Separate areas 1, 2, and 3	
26–28	Oöid A/C sections vs. bedding	
29 a and b,	Lineation in greenstone	192
c–e	Lineation in greenstone	103
30 a and b	Lineation in gneiss	92
31–47	Slickenside diagrams; see Table 9	
	Skyline Drive striae	2,178
	Skyline Drive surfaces	1,656
48 a–d	Massanutten Mountain (see Table 10)	
	striae	205
	surfaces	180
49 a–e	5 single locations in Massanutten Mountain	193
50 a–f	6 locations in limestone valley	303
51	Harpers Ferry striae	169
	surfaces	107
53 a	Chilhowee Formation striae	92
53 b	Antietam Formation striae	66
54 a	Growth actinolite	184
c	Growth quartz	105

In Area 1 all observations are from normal upper limbs of folds in which oöid axes are at a large angle to bedding. In Figure 26, 14 AC plots were used because they were portions of sections and not scattered single values.

The geologic map of Rockingham County, Virginia (Brent 1960) shows that Area 1 is an upper limb of an open anticline which is faulted along its crest. On the geological map only southeast dips are shown for bedding.

The three sections of Figure 26 agree well with the geological map. The upper section is in the northeast corner of the Broadway quadrangle west of Mt. Jackson (Fig. 24). Four oölite beds dip southeast at about the same angles. Ratios are very low, and the angle between oöid A and bedding is very large. In the other three sections the angle approaches 90°.

Ellipticity of the oöids is not primary and parallel to bedding but is due to folding. Cleavage cannot be seen at this low ratio, and the orientation of ellipses must be essentially a function of bedding dip and a rather feeble compression almost parallel with bedding. Measurements of such low ratios are difficult and not as precise as at higher ratios.

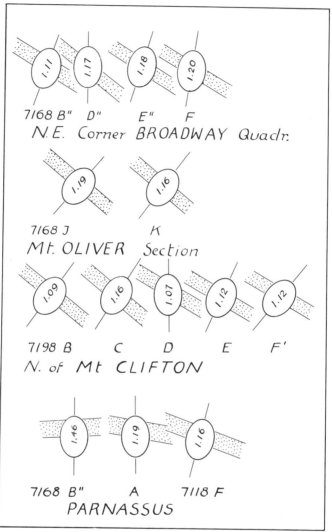

Figure 26. Plot of oöid AC sections in Area 1 for four localities with several oölite beds. Figures in ellipses are A/C ratios.

Figure 27 is a plot of 23 AC sections in Area 2. The lower three sections are in southeast-dipping limbs of a faulted anticline (Brent 1960, map), and the top section is in the northwest-dipping limb of an anticline. The angles between bedding and A are less than in Figure 26, the ratios are much higher, and cleavage can be seen at Edinburgh and most locations, but is not pronounced. If the sections were combined into one fold the cleavage would fan about 45° or less. Overturned limbs with high ratios have not been found, but the Edinburgh section suggests asymmetry, distinct cleavage, and maybe an overturned fold.

Figure 28 is very different, and more than half the measurements in Area 3 are from overturned or vertical limbs. The Shepherdstown section is an exception and includes 11 measurements within a 2-mile radius of

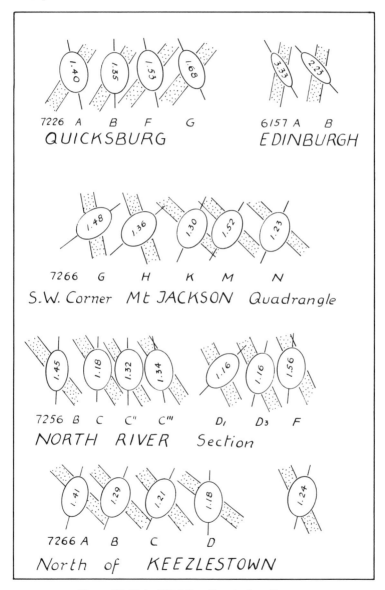

Figure 27. Plot of 23 AC sections in Area 2.

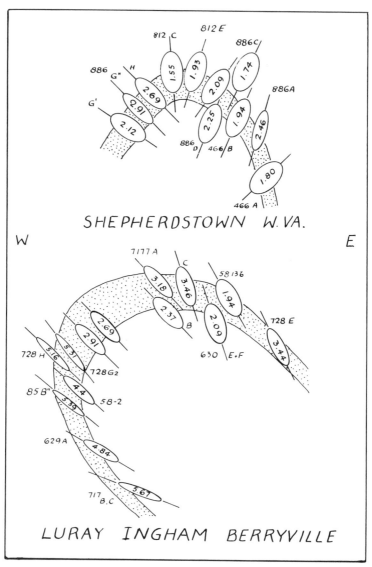

Figure 28. Plot of 25 AC sections in Area 3 for two localities.

the center of town. The combination into one fold may be an artificial construction, but it shows that west-dipping oöid axes occur next to the much more common southeast-dipping ones.

The three samples $886G_1$, G_2, and H may be in the hinge area of an overturned and asymmetrical fold, as shown below, where $886G_2$ and H have also been used. This would not change the fact, however, that samples 886A, C, and D are in the east limb of an upright fold with fanning cleavage.

More prevalent are the orientations of the lower section in Figure 28. Here an asymmetrical, overturned fold has been reconstructed from three fragmental folds at Luray, Ingham, and Berryville. In all of these, cleavage is very intense, ratios are high, and A dips southeast in AB = ab and parallel with the lineation elsewhere. It would not be possible to combine most of the values with those in Areas 1 and 2 or with the eastern half of the fold at Shepherdstown.

Ramsay (1967, p. 343, Fig. 7–1) shows the progressive development of a fold and notes that:

> The state of finite strain in the material varies from point to point. Folded rock strata (formed from initially plane-parallel stratification) are always the result of inhomogeneous finite strain.... If there are markers of known original shape in the deformed rocks, we can completely determine the states of strain throughout the folded material. A classification of folds from this knowledge would be unambiguous and complete, but unfortunately such information is seldom available in naturally deformed rocks, so the method is not practicable.
>
> Even if the finite-strain states throughout the structure are known at every point within a fold, there are an infinite number of possible intermediate deformation paths linking the undeformed initial state to that in a final fold (although not all of these possible deformation paths are equally likely to occur). It is therefore impossible to determine with precision the mechanism of formation of the fold, as we have no knowledge of the strain increments.

In spite of this rather sobering appraisal, it would seem very much worth while to find completely exposed folds with several oölite beds and to determine the ellipsoid axes and their orientations in relation to all parts of the folds and cleavage and lineations.

Even with the meager data at hand, some worthwhile conclusions are possible: In the three oölite belts deformation intensities (A/C ratios) increase from northwest to southeast, and the orientations become asymmetrical approaching the Blue Ridge. The decrease of the angle between cleavage and bedding must mean less fanning of the cleavage with greater deformation intensity. At this stage we do not know whether a cleavage that fanned through about 90° has been rotated during more intense deformation or formed at a low angle with bedding under more intense compression than is reflected in the A/C ratios.

Regionally the oöid A axes occur in a girdle together with cleavage and bedding poles. The girdle is normal to the average for fold axes and is the major ac plane of the deformation plan which dominates the tectonic evolution of that portion of the Appalachians between the northwest end of the Blue Ridge and Lexington, Virginia. The oöid A axes are in the same ac girdle as the lineations in all formations from the Precambrian to the Devonian and are thus witness to the participation of all elements in the over-all Blue Ridge–South Mountain plan. They may vary in limbs of individual folds but remain spokes in the ac "wheel."

Lineation patterns

Penetrating lineations (A, B, C; Table 5)

In the Catoctin Volcanics lineation was plotted in two groups which were taken from two sets of notebooks in order to avoid deliberate selection. Figure 29a shows 192 points in equal-area projection, and Figure 29b is the graphical presentation of 192 points in zones across the diagram of Figure 29a. Almost all points dip southeast, and the maximum direction is 120°. In Figure 29c and d, 103 measurements were plotted. The maximum is at 115°. That diagram has been contoured and is shown as Figure 29e.

Lineation in the gneiss is as well oriented as that in the volcanics but is not as common. The more general orientation in the gneiss parallels that of the volcanics, and the maxima of Figure 30a are at about 115°.

In the overlying sedimentary rocks the lineation listed in Table 5 under B continues upward through the section without deflection through the Chilhowee rocks, where it is very clearly seen in the Loudoun and Harpers Formations and in the orientation of oöid axes.

Nonpenetrating lineations

Lineations D, E, and F (Table 5) occur on surfaces like cleavage or fractures and are not due to arrangement of components within the rock.

In Figures 31 to 47 the orientations of striae and slickensided surfaces have been plotted for the subareas listed in Table 9 for Skyline Drive and some cross roads. In Figure 52 the subareas are shown by their numbers at the locations of the concentrations of points. The number of measured points, planes, contour intervals, contoured diagrams, and rock types are also listed in Table 9. Of 26 maxima only 2 are outside the girdle which includes a zone 20° wide on either side of the center of the diagram. Areas 2 and 8 are in gneiss, and they may suggest a somewhat different orientation of slickensides in gneiss. The center zone, only 10° on either side of the center, is the most densely occupied zone containing all but 5 maxima and the maximum for lineation L (penetrating) in the greenstone of the Skyline Drive, including over 600 measurements.

The maximum *ac* girdle for slickensides and lineations is at 120° azimuth with a preferred dip to the southeast.

Striae are not equally sharply grouped within the *ac* girdle. Figure 35b, for instance, is conspicuously different from Figures 31b, 32b, 37b, 38b, 39b, 40b, 43b, or 44b; Figure 35b shows striae in gneiss, whereas all higher concentrations are in greenstone.

This may be due to the character of the rock, its grain size, its primary textures and structures, absence of a good cleavage in gneiss, or a combination of factors. It may also be due to a tectonic situation in which the

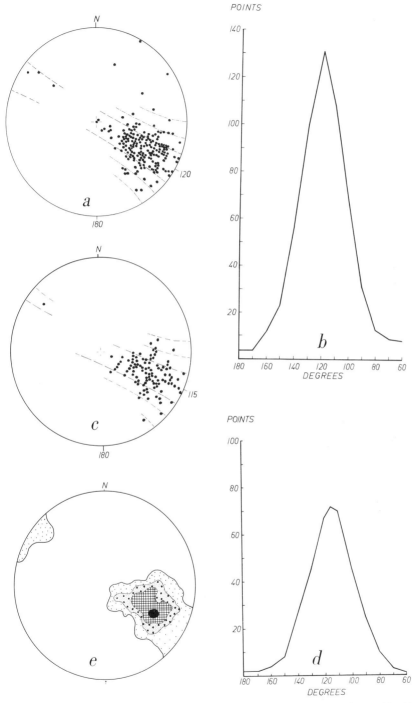

Figure 29. Orientation of lineation in greenstone: (a) 192 points; (b) graph of a; (c) 103 points; (d) graph of c; (e) counted and contoured diagram a. Contours: 1–5–10–17 percent.

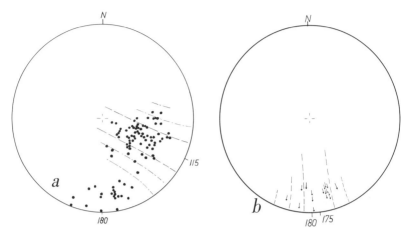

Figure 30. Orientation of lineation in gneiss: (a) 92 points; (b) primary lineation at Old Rag Mountain.

basement below the volcanics has been moved, altered, deformed, and generally strained differently from the relatively thin cover of fine-grained, less massive volcanics.

The orientation of the slickensided surfaces on which striae occur is shown in the c diagrams of Figures 31 to 47. The surfaces are far more scattered and are well oriented only in Figures 32, 34, 38, 40, 43, 44, and 47. The peaks of concentrations need not coincide with those of the striae.

In the gneiss (Fig. 35b) the peak of slickensides occurs at about 143 and for striae at 100. The gneiss subarea south of Swift Run Gap shows the gneiss girdle at 150 and the striae at 130. There is not enough information on this deviation to justify drawing conclusions, but it could mean that pre-existing surfaces in the gneiss became slickensided, whereas the close relation in the greenstone may mean formation of new surfaces (cleavage) and striation during the same tectonic phase or process.

Additional investigation may well show relationships between bedded sedimentary sequences, massive basement, intermediate volcanics, and the availability of surfaces which are now slickensided. It may also be necessary to separate types of slickensided surfaces in the field far more rigorously than has been done so far.

Western slope and valley

Within the circle of Figure 52 are the maxima for Blue Ridge diagrams, and outside for comparison are the maximum girdle (ac) azimuths for the localities listed in Table 10 and shown in Figures 48 to 51.

Figure 48 a–d includes all measurements within Massanutten Syncline between the north and south ends of Massanutten Mountain. Measure-

Table 9. Key to diagrams of slickensides and striae on Skyline Drive,
Shenandoah National Park

Figure	Subarea	No. of Measurements:		Rock type
		Striae	Planes	
31	01	108	106	Greenstone
32	01–02	181	163	Greenstone
33	03	73		Greenstone
34	1	105	111	Greenstone
35	2	78	75	Gneiss
36	3	116	—	Greenstone
37	4	163	141	Greenstone
38	4A	104	109	Greenstone
39	5	230	177	Greenstone
40	6	103	80	Greenstone
41	7	83	72	Greenstone
42	8	175	220	Gneiss
43	8A	186	183	Greenstone
44	9, 10, 10A	163	161	Chilhowee
45	11A	106	117	Greenstone
46	11	102	—	Greenstone
47	12	102	81	Greenstone

Table 10. Slickensides and striae—single localities

Figure	Locality	No. of measurements		ac girdle azimuth degrees
		Striae	Planes	
48 a–d	Massanutten Mtn.	205	180	132
49a	Catherine Furnace	26	—	100
49b	Brocks Gap	48	—	130
49c	Woodstock Gap	29	—	140
49d	Edinburg Gap	74	—	125
49e	East of Woodstock Gap	16	—	130
	Limestone Valley			
50a	Luray area	41	—	115
50b	Front Royal area	150	—	105
50c	Charles Town–Shepherdstown area	23	—	115
50d	Ingham area	24	—	130
50e	Berryville area	8	—	120
50f	Alma area	57	—	120
	Formations			
53a	Chilhowee	92		120
53b	Antietam	66		105
	Growths			
54a	Actinolite	184		120
54b	Quartz	105		130

ments are from slickensided surfaces that could be found on accessible roads and including all formations from Martinsburg to the Romney shale. The surfaces (Fig. 48c) are mostly steep and trend northeast and northwest, with gradations but with no horizontal surfaces and few gently-dipping ones. Correspondingly, striations are steep and concentrate in the center, but with considerable scatter. The maxima in Figure 48b for striae and poles of planes coincide with an azimuth for the zones of 130. The *ac* girdle is pronounced.

Five localities in Massanutten Syncline are separately shown in Figures 49 a–e. The number of slickensides fluctuates between 16 and 74, depending on outcrops and on formations. In the Tuscarora Sandstone, slickensides are widely spaced and therefore not frequently seen; they are also more readily weathered than in Martinsburg Shale, which has a much lower general porosity. It is also worth noting that the *ac* girdle at Catherine Furnace shows an azimuth of 100, whereas all other localities are between 125 and 140.

It is a tempting project to systematically map slickensides in the Massanutten Syncline for a number of key formations with different competences and to include the Martinsburg Shale to the north and south for at least 20 miles along strike. The data now available are barely a sample of what can be expected.

Figure 50 a–f is a group of six localities in the limestone valley along the Blue Ridge foothills. In diagrams 50 a, b, c, and e lineations are also shown as arrows, some oöid A as black ellipses, and cleavage poles as crosses to provide information on mutual relationships between structure elements. The zones were constructed for striae densities. Their azimuths are shown on Figure 52 as black spots outside of the circle.

The Potomac River gorge at Harpers Ferry is shown in Figures 51 a–d. The surfaces in Figure 51c do not culminate directly below the peak of striations, which is very large. This may be due to the presence of the Weverton Formation with its massive quartzites and slickensided joint systems. A more detailed collection of data may well establish the reasons for the deflection. Surfaces also show more scatter than striae, confirming the observation that striae are far more consistent than the surfaces on which they occur. The striae maximum (Fig. 51b) is exactly in the average girdle for the Blue Ridge striations. Slickensides in the large South Mountain cut, east of Harpers Ferry on U.S. 340 (Fig. 51a), are in a zone which differs only 5° from that in the Elk Ridge cut to the west.

Growths in fractures

Growth of actinolite, quartz, and calcite indicates also a surprisingly constant direction which coincides with slickensides and lineation (Fig. 54). Figure 54 shows the plot of 184 actinolite needles and graphical presentation of orientation. Figure 54 c and d are the plot and graphical presentation of 105 quartz needles. There is some scatter in Figures 54a and c, which must be expected where minerals grow in fractures that are not necessarily well aligned. The zones show distinct maxima in Figures 54b and d, which leave little doubt about the orientations. The difference between an ac girdle orientation of 120 and 130 is not significant. It is, however, surprising that there is a consistent orientation over a large area.

Summary on orientations

The orientations of structure elements were plotted into equal-area nets and are presented in the figures listed in Tables 8–10. In addition, the orientation of a lineations, cleavage, and oöids A axes are shown in Plates 1 and 2. Orientations of slickensides, striations, lineation, and oöid A axes are shown in Figure 52. The slickenside ac girdles of maximum numbers of points are also shown in Figure 55. Finally, Figure 56 summarizes the mutual relationships of fold axes, cleavage-bedding intersections, lineation, bedding-pole girdle, and the slickenside zone of Figure 52. Over a wide area and with elimination of many local deviations the plan is as follows:

Fold axes and S_0–S_1 intersections trend 35° and are about horizontal, defining a regional direction b. Normal to this b is an ac girdle which is occupied by a bedding-pole girdle with two maxima. Girdle trend is 125. The maximum for lineation and cleavage in the volcanics is also in this girdle and almost coincides with the slickensides maxima of the volcanics (Fig. 52). Oölite axes girdle is at 130 or 5° clockwise from fold axes-ac girdle and 10° from slickenside maximum. The deviation is small and within the limit of error of single measurements and projection-net manipulations. With the large number of measurements it could be significant even if very small. There is a systematic rotation of direction in Figure 55 from 115 to 120 to 130 in the southern end of the Blue Ridge near Waynesboro.

As thrusting becomes more prevalent southward with the Pulaski thrust reaching toward the south end of Massanutten Mountain and the Appalachian trend changing at Roanoke, it may be that this change is real. A continuation of this study toward Roanoke could readily determine a possible change in trend.

As it is, the average ac plane is vertical with horizontal fold axes, a strong lineation dips southeast also normal to b and in the cleavage plane ab and the ac plane normal to b. The symmetry is monoclinic, except for small subareas, and is overturned to the northwest. The symmetry plane is ac.

Rotation is to be expected because horizontal beds cannot be transferred into vertical or overturned ones without rotation. Even on the small scale of a thin section there are many indications of rotation.

Table 11. Structures in key units that can be used in determining
deformation plan

Devonian	Romney	Cleavage
		Slickensides and striae
		Mineral growths
		Fold axes
Silurian	Tuscarora	Slickensides
		Mineral growths
Ordovician	Martinsburg	Cleavage
		Slickensides
		Mineral growths
		Flattened crinoids
		Fold axes
		Cleavage-bedding intersections
	Athens	Cleavage
		Lineation b
		Slickensides and striae
		Striae on bedding
		Mineral growths
		Fold axes
		Bedding—cleavage intersections
	Beekmantown	Cleavage
		Lineations b and a
		Slickensides and striae
		Mineral growths
		Boudinage
		Oölite ellipsoids
		Bedding–cleavage intersections
Cambrian	Conococheague	Cleavage
		Slickensides and striae
		Oölite ellipses
		Boudinage
		Crenulations
		Lineations a and b
		Fold axes
		Cleavage–bedding intersections
	Elbrook	Cleavage
		Slickensides and striae
		Oölite ellipses
		Boudinage
		Crenulations
		Lineations a and b
		Fold axes
		Cleavage–bedding intersections
Chilhowee	Antietam	Slickensides and striae
		Mineral growths
		Scolithus
		Lineation, rare
	Harpers	Cleavage 1 and 2
		Lineations a and b
		Slickensides and striae
		Bedding–cleavage intersections
		Crenulations

Table 11.—*Continued*

	Weverton	Cleavage
		Lineations *a* and *b*
		Slickensides and striae
		Growths
		Intersections
		Recrystallization
Precambrian	Catoctin Volcanics	Cleavage
		Lineations
		blebs *a*
		crenulations
		Slickensides and striae
		Growths, fibers
		Amygdules
		Blebs
	Gneiss	Cleavage
		Lineations
		Shear zones
		Blebs
		Growths, fibers
		Slickensides and striae

This simple pattern is universal in all diagrams and all formations from the Precambrian gneisses to the Silurian. The deformation plan is identical with that reported from South Mountain, Maryland (Cloos 1947). Deviations are only a matter of orientation in space, but not of mutual relations of the structures which are the basis of choosing *a*, *b*, and *c*. The elements are common everywhere, can be identified in the field, in specimen, and thin sections. There can be little confusion if the co-ordinates can be so easily recognized.

Even if the co-ordinates are mutually consistent their geographic position may vary from place to place, because folds plunge, trends change, faults interfere, and overthrusts eliminate portions of the stratigraphic column and some of the structures.

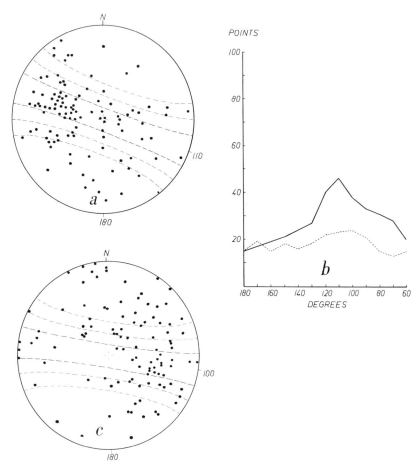

Figure 31. Striae and slickensides, Subarea 01, Skyline Drive, mile 0–10. Greenstone: (a) 108 striae; (b) graph of a and c. Solid: striae; dotted: poles of surfaces; (c) poles of surfaces.

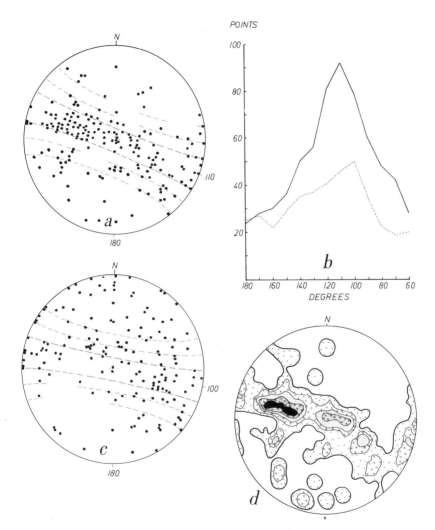

Figure 32. Striae and slickensides, Subareas 01–02, mile 0–19.3. Greenstone: (a) 181 striae; (b) graph of a and c. Solid: striae; dotted: surfaces; (c) poles of surfaces; (d) contoured diagram a. Contours: 1–2–3–5–6 percent.

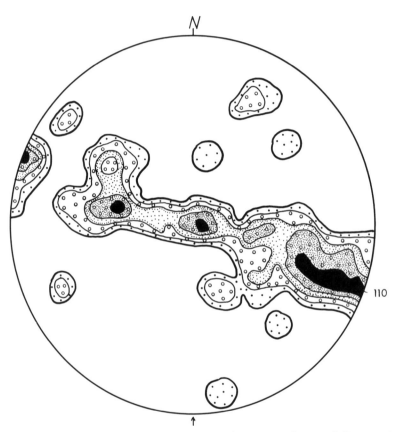

Figure 33. Striae subarea 03. Skyline Drive, miles 21.5–25. Contoured diagram of 73 points. Contours: 1–2–3–4–5–6–7 percent.

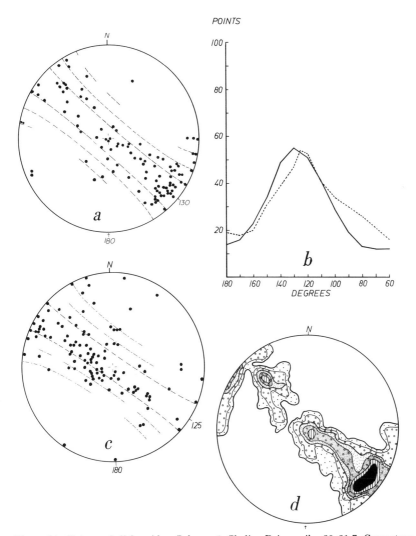

Figure 34. Striae and slickensides, Subarea 1. Skyline Drive, miles 30–31.7. Greenstone between Chilhowee Formations and Thornton Gap: (a) 105 striae; (b) graph of a and c. Solid: striae; dotted: poles of slickensides; (c) 111 poles of slickensides; (d) contoured diagram a. Contours: 1–2–3–5–7–(8–11) percent.

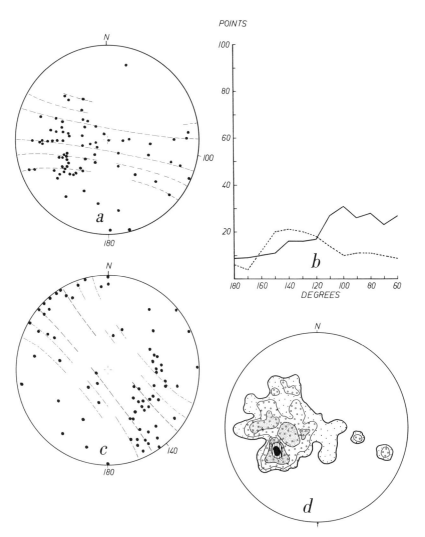

Figure 35. Striae and slickensides, Subarea 2. Skyline Drive. Thornton Gap to Stony Man, miles 31.7–37. Gneiss: (a) 78 striae; (b) graph of a and c. Solid: striae; dotted: surfaces; (c) 75 poles of slickensided surfaces; (d) contoured diagram a. Contours: 1–2–4–6–(8–11) percent. Note the larger scatter of surfaces as compared to greenstone.

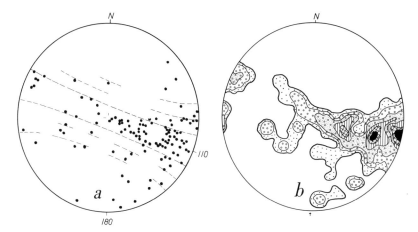

Figure 36. Striae, Subarea 3. Skyline Drive, Stony Man to Crescent, miles 39–44. Greenstone: (a) 116 striae; (b) contoured diagram a. Contours: 1–2–4–8–10–12 percent.

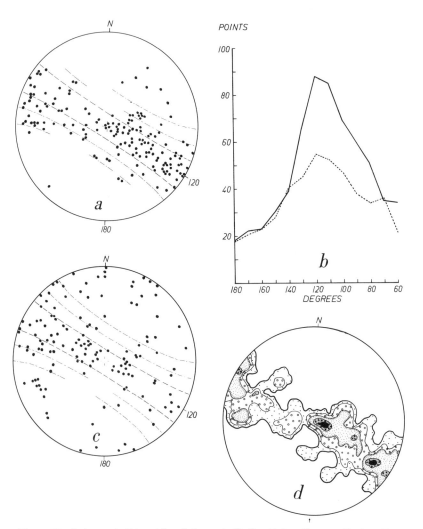

Figure 37. Striae and slickensides, Subarea 4, Skyline Drive. Crescent Rock to Tanners Ridge, miles 44–52.5. Greenstone: (a) 163 striae; (b) graph of a and c; (c) 141 poles of slickensides, (d) contoured diagram a. Contours: 1–2–3–4–(6–7) percent.

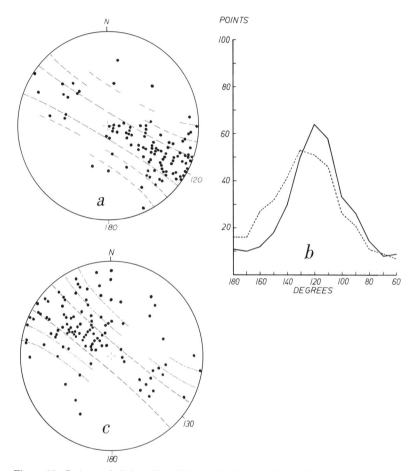

Figure 38. Striae and slickensides, Subarea 4A. Skyline Drive, Rapidan Fire Road east of Big Meadows. Greenstone: (a) 104 striae; (b) graph of a and c; (c) 109 poles of surfaces.

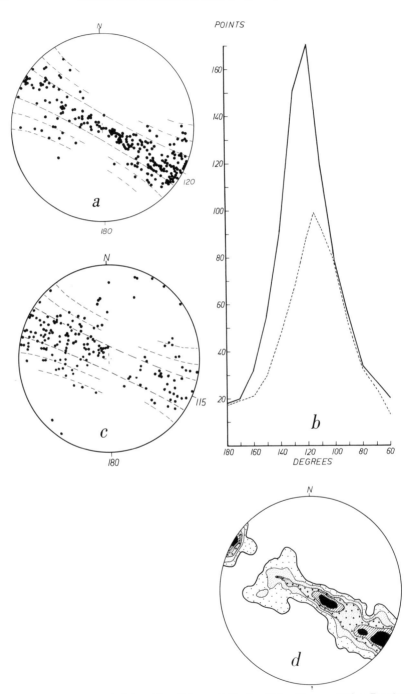

Figure 39. Striae and slickensides, Subarea 5. Skyline Drive. Tanners Ridge–Bearfence Mountain, miles 52.5–56. Greenstone: (a) 230 striae; (b) graph of a and c; (c) 177 poles of surfaces; (d) contoured diagram a. Contours: 1–3–5–7–(11–13) percent.

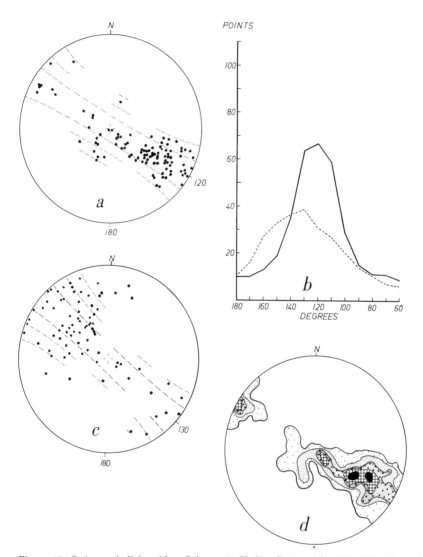

Figure 40. Striae and slickensides, Subarea 6. Skyline Drive, miles 56.5–62.6. Green-stone: (a) 103 striae; (b) graph of a and c; (c) 80 poles of surfaces; (d) contoured diagram a. Contours 1–3–5–7–9–(11–12) percent.

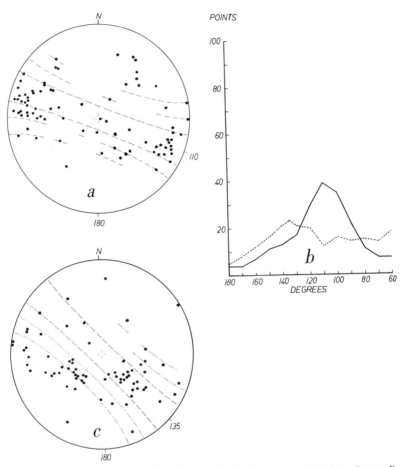

Figure 41. Striae and slickensides, Subarea 7. Skyline Drive. Swift Run Gap to Park boundary on road to Elkton. Greenstone: (a) 83 striae; (b) graph of a and c; (c) 72 poles of surfaces.

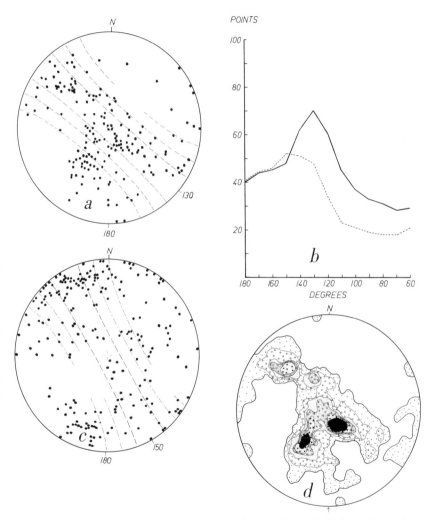

Figure 42. Striae and slickensides, Subarea 8. Skyline Drive. Swift Run Gap south to mile 70.3. Gneiss: (a) 175 striae; (b) graph of a and c; (c) 220 poles of surfaces; (d) contoured diagram a. Contours 1–2–3–4–5–6 percent.

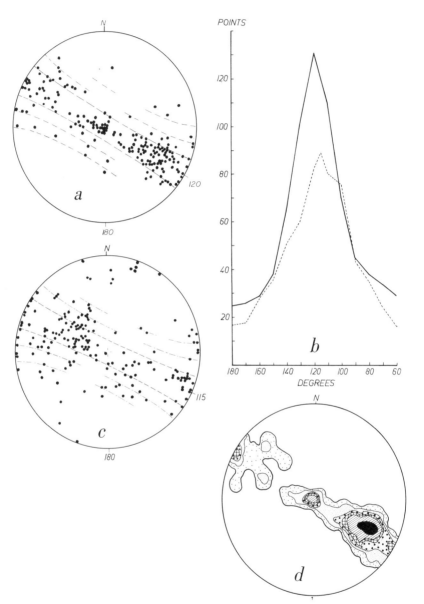

Figure 43. Striae and slickensides, Subarea 8A. Skyline Drive, miles 70.3–75. Green-stone between Gap and Chilhowee formations: (a) 186 striae; (b) graph of a and c; (c) 183 poles of surfaces; (d) contoured diagram a. Contours: 1–3–5–7–9–11 percent.

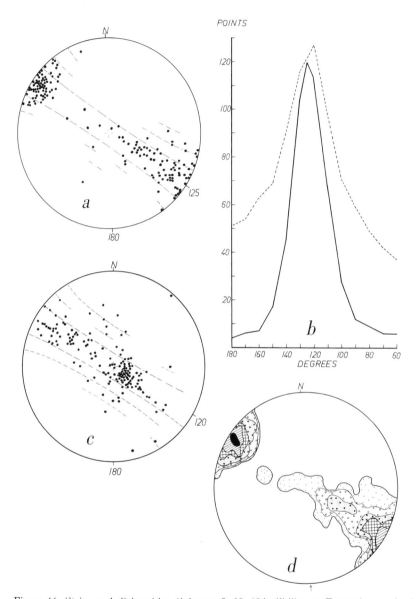

Figure 44. Striae and slickensides, Subareas 9, 10, 10A. Chilhowee Formation south of Swift Run Gap, miles 75–96. (a) 163 striae; (b) graph of a and c. Solid: striae; dotted: poles of surfaces; (c) 161 poles of surfaces; (d) contoured diagram a. Contours: 1–2–5–7–9–(11–20) percent.

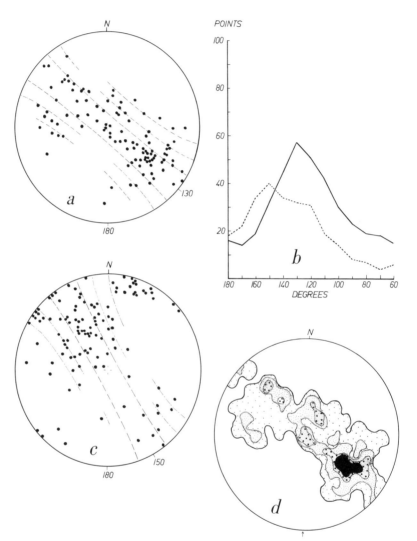

Figure 45. Striae and slickensides, Subarea 11A, north of Rockfish Gap. Greenstone along Skyline Drive: (a) 106 striae; (b) graph of a and c. Solid: striae; dotted: striated surfaces; (c) 117 poles of surfaces; (d) contoured diagram a. Contours: 1–3–5–(7–14).

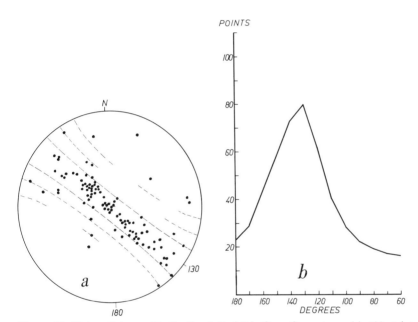

Figure 46. Striae, Subarea 11. South of Rockfish Gap. Greenstone: (a) 102 striae; (b) graph of a.

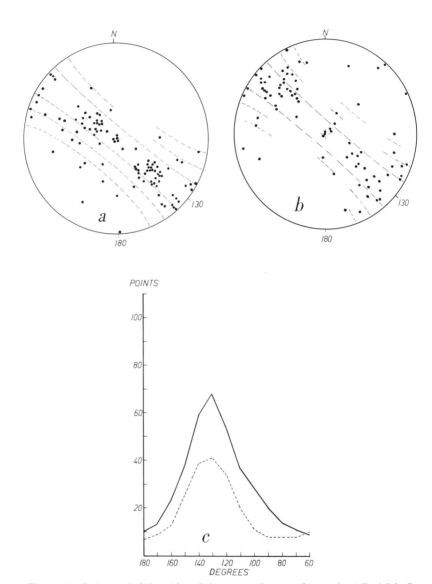

Figure 47. Striae and slickensides, Subarea 12. Seven miles south of Rockfish Gap. Greenstone: (a) 102 striae; (b) 81 poles of slickensided surfaces; (c) graph of a and b. Solid: striae; dotted: surfaces.

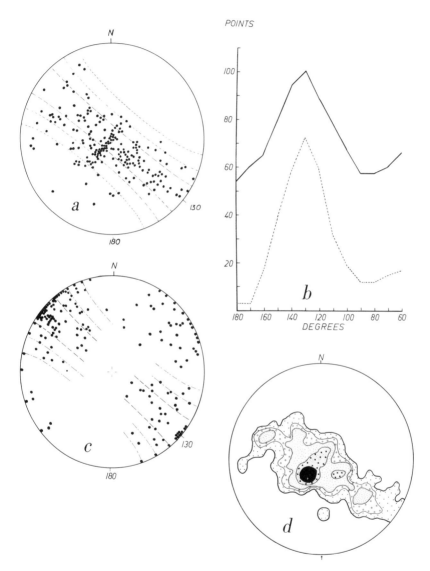

Figure 48. Striae and slickensides. Massanutten Mountain: (a) 205 striae; (b) graph of a and c. Solid: striae; dotted: striated surfaces; (c) 180 poles of surfaces; (d) contoured diagram. Contours: 1–2–3–4–5–10 percent.

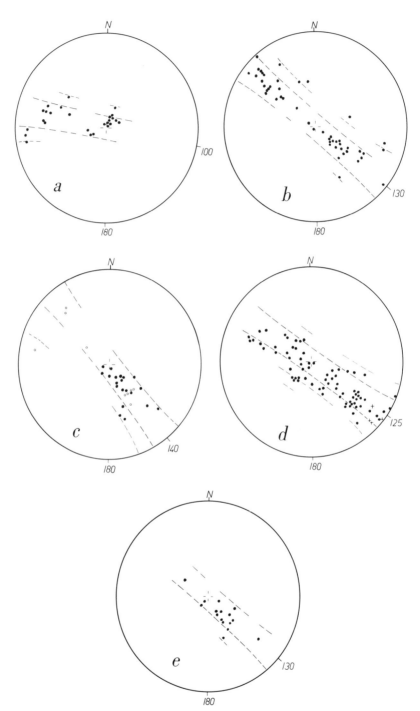

Figure 49. Striae, single localities, Massanutten Mountain: (a) Catherine Furnace, 26 striae; (b) Brocks Gap, 48 striae; (c) Woodstock Gap, 29 striae; (d) Edinburg Gap, 74 striae; (e) east of Woodstock Gap, 16 striae.

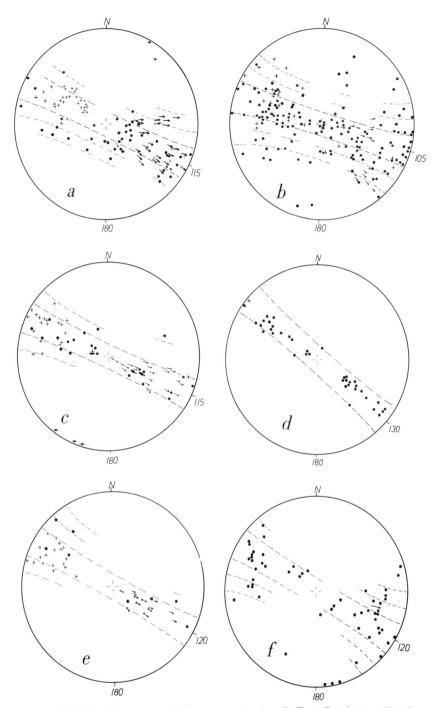

Figure 50. Striae in limestone: (a) Luray area, 41 striae; (b) Front Royal area, 150 striae; (c) Charles Town-Shepherdstown area, 23 striae; (d) Ingham area, 24 striae; (e) Berryville area, 8 striae; (f) Alma area, 57 striae. Arrows show lineation; crosses are cleavage poles.

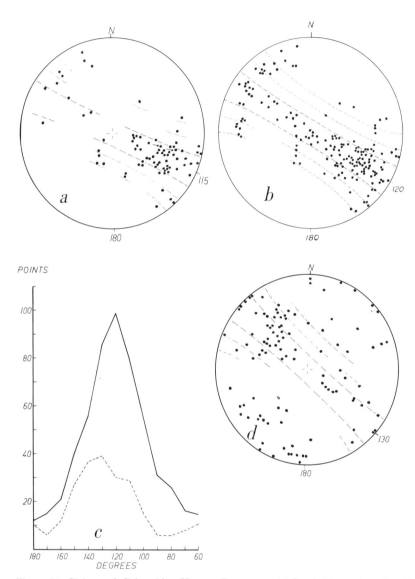

Figure 51. Striae and slickensides, Harpers Ferry area. (a) South Mountain striae; (b) 169 River Gorge; (c) graph of b and d; (d) 107 poles of striated planes.

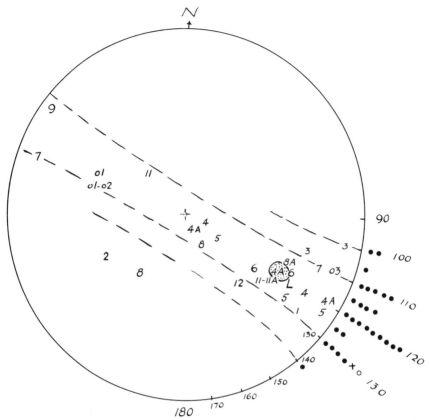

Figure 52. Synoptic diagram of striae maxima of Figures 29–51. Subareas numbered. L: lineation maximum for all measurements. Black dots outside circle are maximum azimuth readings of striation diagrams. 03–12: subareas along Skyline Drive. X: azimuth for oölite axes. Open circle: azimuth for lineation in limestones.

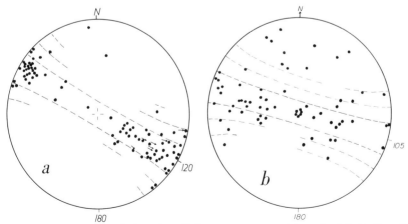

Figure 53. Striae and slickensides, Chilhowee Formation: (a) south of Swift Run, 92 striae; (b) Antietam Formation, 66 striae.

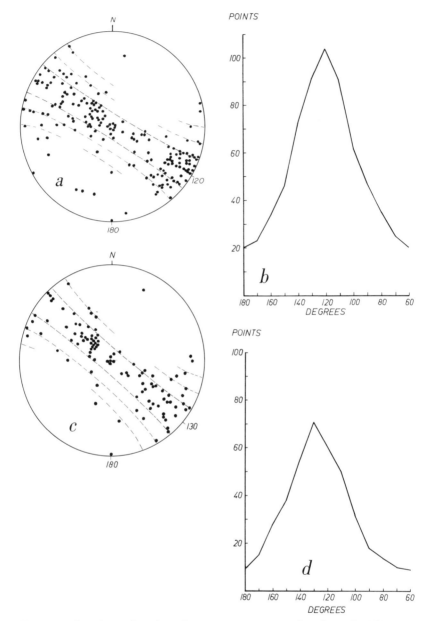

Figure 54. Growth of rods and needles: (a) 184 actinolite needles; (b) graph of diagram a; (c) 105 quartz rods; (d) graph of diagram c.

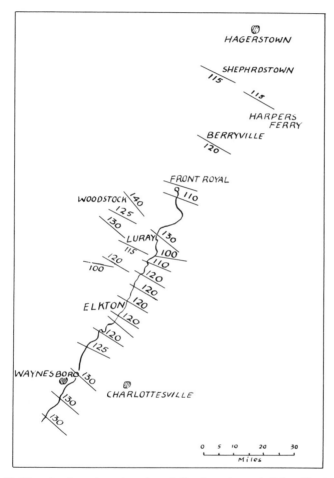

Figure 55. Map showing orientation of *ac* girdles for striae and slickensides in Area 3

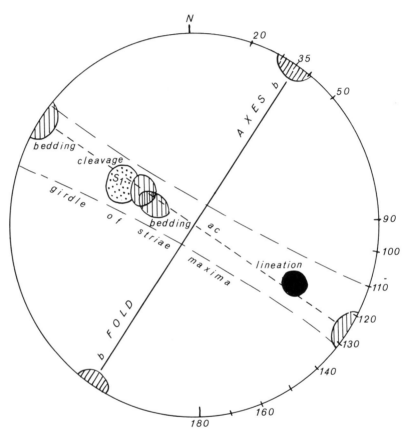

Figure 56. Synoptic diagram showing orientation of maxima for bedding poles, cleavage poles, lineation, girdle for striae, and fold axes.

XII

Summary

This study is an extension of my 1947 oölite project in several directions: it includes a much larger area to the south, almost to the James River; it presents improved measuring techniques, a more detailed description and analysis of oölites and associations, the effect of deformation on different species, and data on the deformation mechanism of spherulites; deformation intensities were mapped, and lineation, cleavage, slickensiding, and mineral growths in fractures from the Precambrian to the Devonian are included. There are also observations on formation of cleavage in relation to strain. (For raw data, see appendix, p. 207.)

The extension over such a large area was possible only because a consistent deformation plan exists and its orientation can be determined with proper sampling and the use of any of the structure elements which exist at a given locality or in a formation. Oölites in limestone, lineation and cleavage in volcanics and gneiss, slickensides in the Martinsburg, or cleavage-bedding intersections in the Harpers Formation are equally useful if measured in large numbers and at many localities. They are not equally useful in the quantitative determination of strain, but can be used qualitatively in the analysis of strain gradients, their distribution and orientations.

The South Mountain plan has been confirmed in the enlarged area. More complicated geological situations do not change the mutual relations of structure elements. The group of thrusts at Front Royal did not influence the deformation plan because it is not a function of movements of blocks, but of penetrative deformation. The deformation of oölites or volcanic blebs and the formation of cleavage and lineation are due to "rock flowage."

The consistency of the deformation plan is demonstrated in Plates 1 and 2 and in the diagrams.

We measured more oöids per slide as species were more clearly defined and as we found more oölites with feeble orientations in Area 1. Many saw cuts were measured with binoculars, and this can be more satisfactory in some samples and increases the number of cuts that can be analyzed. It also saves time and should be standard procedure for the determination of the principal oöid axes and planes. Thus accuracy was increased, time was saved, and more has been accomplished with the funds available.

The distinction between oöid species is essential, and in mapping strain gradients only like species should be compared. Most widely distributed and most useful are spherulites and layered oölites with spherulite centers. Least useful are mud pellets, except as indicators of directions. Not usable are dolomite clusters, pseudoölites, and quartz. Elimination of freaks is necessary also because fossil fragments with oölitic growths never were spherical. It is also worth while to separate sizes. Most desirable is a separation of types, which can be combined later but can never be separated if they are not observed separately first.

Because they freeze a state of deformation and permit observations on the history of deformation within one sample, chert nodules are useful. In addition, the sense of rotation can be observed where chert nodules are rotated.

Deformation intensities increase from northwest to southeast. The average A/C ratio in Area 1 is 1.16, in Area 2 it is 1.34, and in Area 3, along the Blue Ridge foothills, it is 2.20. The highest values are 1.46, 1.72, and over 6.00. Along strike the values also vary, depending on the position of oöids in folds and large tectonic units.

Other strain markers exist, but their accuracy is not comparable with that of oölites. In the Blue Ridge volcanics there are chlorite blebs whose A/C ratios exceed 100, but their origin and original shape is unknown.

Oölite deformation is partly ductile flow, but greatly aided by fracturing and coalescence of spherulite fibers into crystal wedges. Cleavage appears at A/C ratios above 1.5 and becomes visible as mineral orientations and parallelism of components.

Lineations are either parallel to fold axes, seen as intersections of cleavage and bedding and as crenulations, or they are normal to fold axes in the prevailing flow cleavage (ab). It is represented as oöid A axes, mineral alignment in sediments, chlorite blebs and fibrous mineral growths in volcanics, and as spindle-shaped bodies or streaks in the gneiss. In the gneiss the lineation is partly primary (Old Rag Mountain), and orientation is not as rigid.

Striations on slickensides and cleavage poles are in one girdle with cleavage and bedding poles and lineation (a), but not always as well defined and rigorously oriented. This is not surprising because striae are on joints which formed toward the end of deformation. Slickensides and striations are most pronounced in the Catoctin Volcanics, but they occur in the same orientation pattern in Massanutten Syncline, the limestones, Chilhowee formations, and the gneiss.

Even the secondary fibrous mineral growths of quartz, actinolite, and chlorite in fractures are oriented predominantly in the ac girdle.

The orientation is constant. The decrease in intensity is gradual and affects the same formations. Structures are the same in equivalent formation. Cleavage and lineation (*a*) cut across volcanics, the Chilhowee, and the limestones. Tectonites grade into nontectonites at about the same distance from the crest of South Mountain–Blue Ridge. Folds are asymmetrically overturned west, cleavage dips east, and symmetry is monoclinic.

The stratigraphic thickness in which the deformation plan exists is at least 20,000 feet. It includes several thousand feet of Precambrian gneiss and volcanics, the Chilhowee Group, and the limestones. Not all structure elements occur in all rocks, but the orientations are represented by several structure elements. Chlorite blebs indicate lineation in the Catoctin Volcanics. Oölite A axes serve the same purpose in the limestones. Cleavage occurs from the gneiss to the Tuscarora, and slickensides are everywhere.

Thrusts help to telescope intensity zones at Front Royal and to the south where they are more numerous and larger than in the Central Appalachians.

XIII

Conclusions and Discussion

The consistent and incredibly uniform orientation of the deformation plan is limited to a zone about 30 miles wide and 200 miles along strike from Pennsylvania to the James River, Virginia. It includes part of the crystallines and the sedimentary rocks of the adjacent Great Valley. Vertically it includes basement gneiss, Catoctin Volcanics, and Paleozoic sediments up to the Silurian and, possibly, the Devonian. In this zone the surface of the Precambrian basement and floor of the Appalachian Geosyncline rises abruptly to the surface and permits study of the relationship between basement and geosynclinal fill. To the west, the basement disappears. To the east, all rocks are metamorphosed, highly altered basement, possibly with remnants of geosynclinal rocks and large masses of controversial crystallines such as the Wissahickon Schist in Maryland. The contact between basement and Paleozoic rocks is not a sharp tectonic break but a tectonic transition. This does not mean that it cannot be sharp to the west or that the basement is involved in Applachian folding to the west of the transitional zone.

Rodgers (1949) reviewed the evolution of thought on Appalachian structure and developed ideas on Appalachian folding. One of the fundamental problems is the support of the foreland and Ridge and Valley areas by a more or less rigid basement during the folding. The question was asked "Does it or does it not involve the basement?" (Rodgers 1949, p. 1651) and thus the expressions "thin-skinned" and "thick-skinned" tectonics. The answer is probably not a simple "either/or" but is more complicated. Where we see it, the basement is intensely involved.

Circumstantial evidence suggests that the basement cannot participate as fully below the Ridge and Valley and foreland as in the Blue Ridge because folding of horizontal strata increases their thickness—17,000 feet of stratigraphy becomes 34,000 feet in a reduction to half of the width (Cloos 1940). If the basement participates equally it will contribute to the thickening so that 10,000 feet of basement becomes 20,000 feet after lateral shortening. The thickened basement must be added to the thickened sediments, and the total cannot accumulate into a welt that is too high because isostatic equilibrium would not tolerate it. Even the first premise—the lateral basement reduction—seems questionable. Thus the cover must separate from the basement. If the basement were to carry a passive column and were to disappear into the depth the problem could be solved as suggested by Ampferer (1906), by *Verschluckung* or descending convection currents, presently called "subduction."

Rodgers (1964, p. 74) chose the very tempting solution of thin-skinned tectonics and selectively limited basement participation. The thin skin moves by gravity, and the basement is only slightly involved.

Cooper (1964) points out that sedimentation within the Appalachian Geosyncline is determined by existing large-order anticlines and synclines, and he distinguishes anticlinal and synclinal facies. The local patterns of stratigraphic variation indicate that major Appalachian folds originated during Paleozoic deposition and grew to a large extent as the result of differential vertical downwarp. In spite of subsequent horizontal displacements along some thrusts, the prevailing stratigraphic and structural features suggest that differential vertical movements in the accumulating sediments and supporting basement were dominant during Paleozoic sedimentation and continued to act even after thrusting.

The basement was obviously involved in epirogenetic vertical movements below the Appalachian ridges and valleys. Thrusting on bedding planes and low-angle wedges preceded vertical movements. However, gravity sliding is also a limited mechanism that would be unlikely to result in long straight folds which can be traced for 200 miles. Somewhere there would be bulges and recessions.

In the southern Appalachians, gravity creep on thrust surfaces has moved large sheets forward but not made long and persistent folds. Also, if gravity were the only mechanism the sliding sedimentary cover should tear loose at the rear and form a Bergschrund and border graben as in the Gulf Coast (Cloos 1968). Finally, gravity decollement would leave an undisturbed basement behind, without manifestations of metamorphism such as cleavage, oölite deformation, and lineations that cut across the Cambrian–Precambrian boundary and high into the limestone and deeply into the gneiss.

The surfaces of the Precambrian volcanics and gneiss are now vertical or overturned along the edge of the Blue Ridge. The basement has been metamorphosed, along with its cover, and the westward movement has been substantial. At some places thrusts have increased the lateral forward displacement.

This has also been recognized by Rodgers, who likens the Blue Ridge basement at that time to "plastic goo" (1964, pp. 72, 79). He also recognized the possibility that the basement is not equally involved and that there is not equal consistency between the crystalline axis and the Allegheny front.

If the upturned surface of the Precambrian is returned to its horizontal position, several questions must be answered and some assumptions made: What is the present configuration of the upturned surface? How was it distorted? How was it rotated into its present position? The transfer has been accompanied by vertical thickening and lateral shortening.

Figure 57 is a collection of cross sections that show the crystalline boundary as overturned, near-vertical, with virgation and asymmetry to the west and accordion-like folding. To the west of the crystallines the sediments are folded, repeated, distorted, thickened vertically, and reduced laterally. Cleavage is shown in section A, B, C, E, F, and G. Some structures have been added to the published sections.

The stratigraphic column between Massanutten Mountain and the Blue Ridge is about 17,000 feet between the Silurian and the surface of the Catoctin Volcanics. This sequence is now compressed into about half its horizontal extent or maybe less. This thickened the column, and the basement must have been depressed to maybe 30,000 or 35,000 feet. If the hole west of the Blue Ridge is 35,000 feet deep and the average Precambrian surface is about vertical, flattening of the surface would be like closing a trap door and would place the edge of the crystallines at least 6 miles to the east and below 17,000 feet of sediments. If the Precambrian surface is folded as shown by Nickelsen, the accordion would have to be smoothed out also, and the area that is now the crest of the Blue Ridge would have been at least 8 or 10 miles to the east (Fig. 57, H).

If this deduction is correct the block between Massanutten Mountain and the Blue Ridge has doubled its thickness and reduced its width to half. Since the upturned edge, the crystallines below, and the sediments above are in a zone in which cleavages and lineations cut across contacts from the gneiss to the Silurian the entire complex was deformed together. Any gravity creep and folding must have preceded the upturn of the edge and metamorphism at a pre-cleavage stage.

Pre-cleavage repetition of sections has been described by Cloos and Hietanen (1941) from Lancaster County, Pennsylvania, and has been found by David Elliott (1970, personal communication) in Washington County, Maryland. It may well have occurred geologically above the Blue Ridge and Great Valley and very early in the tectonic history. In the now-preserved section up to the Silurian, gravity creep is unlikely. On the contrary, the section is now welded into a zone of gradationally changing deformation intensities. In addition, the thickening of the folded and upturned section west of the Blue Ridge must have depressed the floor of the trough to greater depths than farther west, where folding is less intense and basement may not be involved. The floor may have sagged isostatically under the increased load of the thickened column after some higher layers, for instance, the Devonian, had crept to the west when the crystallines began to move.

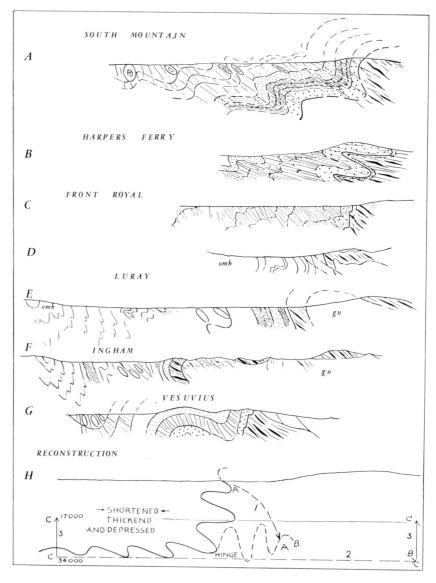

Figure 57. Seven cross sections constructed by: A—Cloos (1947); B—Nickelson (1956); C—Wickham (1969); D—Allen (1967); E—Allen and Cloos (1967); F—Reed and Cloos (1955); and G—Werner (1966). Section H is an attempt at reconstruction of the upturned edge in three stages.

The sequence might have been as follows: activation of the crystallines at depth and rise and westward creep; due to the rise and creep higher layers formed pre-cleavage folds, wedges, and bedding-plane thrusts; formation of folds, cleavage, lineation, metamorphism in the lower layers during westward push from "behind" or from "within" and appearance of steeply east-dipping reverse faults as at Front Royal or those shown by Allen and by Werner (Fig. 57); post-tectonic elevation. Conditions that are critical in the different tectonic zones are tabulated in Table 12.

The steps that have to be taken in a reconstruction of the pre-folding situation are illustrated schematically in Figure 57 H. Stage 1 shows the flattening of the folded and upturned erosion surface. Stage 2 is an attempt to flatten the folds of the erosion surface and of the sediments between Massanutten Mountain and the Blue Ridge. Stage 3 is the restoration of the level of 17,000 feet of sediments up to the Silurian above the erosion surface. Even though the diagram is very schematic it illustrates the enormous amount of mass movement that must be taken into account. If the Devonian is added to the column above the Blue Ridge, the movements must be larger and more complicated. There seems little doubt that the Blue Ridge crystallines have moved considerable distances westward, and upward.

In a simplified way cleavage and lineation suggest the direction of movement and its inclination. It differs in different sections of Figure 57 and is closely related to the formations that are affected. Nickelsen shows recumbent folds, and at Harpers Ferry cleavage is horizontal or dips west; Werner shows cleavage steep in open folds. At Ingham cleavage and oölite axes dip 45° E. in schistose limestone. The cause cannot be gravity

Table 12. Tectonic zones

West of Great Valley	Great Valley and Blue Ridge	East of Blue Ridge
Bedding-plane gravity creep; wedges, repetition of sequence, and thickening by "stacking"	Regional cleavage and lineation; vertical thickening by folding and lateral reduction of width.	Mobilization of Precambrian basement gneiss; shear zones; superimposed cleavages and lineations. Cleavages and lineations in volcanics.
Little over-all lateral reduction at depth; basement probably not involved	Metamorphism increases eastward. Steeply east-dipping thrusts and reverse faults. Upturning of surface of Precambrian basement intensely involved.	Crystallines rise and crowd westward against the depressed sediments.

alone, because gravity creep cannot explain the welding of the lower Paleozoics to the crystallines and their simultaneous deformation in a uniform plan. It can also not explain east-dipping cleavage or east-dipping reverse faults.

Basement participation may have increased gradually and must have been activated from within the crystalline axis of the Appalachians. This could be due to convection, ocean floor spreading, or any other mechanism that will activate the crystallines, move them westward and upward, upturn and fold the Precambrian surface, and produce the multitude of structures that make up a uniform deformation pattern in this large area.

PLATES

Plate 3. Oölite in Conococheague Formation. Large spherulites with concentric layers and radial dust rows. Unattached and single or in clusters of debris or attached to pebbles. Smaller spherulites are half size or less, also layered, some with dolomite. In addition, dolomite grains and clusters, oöid pebbles, mud, and many odd-shaped fossil fragments, one with spherulites inside. Peripheral growth at minimum. Deformation feeble, independent of size. Matrix is sparry calcite with some dolomite. (For details on radial structures see Plates 43, 44). Sample 7256C1. AB cut, south of Harrisonburg, Virginia, Area 2.

Plate 4. Same sample as Plate 3 in plane normal to Plate 3. Sections dip steeply, intersect in almost vertical line. Deformation slight, long ellipsoid axes steep. A/C ratio as determined separately for 100 large and 100 small spherulites 1.18. Dark oöids: (1) very high or low cut; (2) nearer center because it shows layers but black radii are cross cuts; (3) nearer center because it is larger, shows layers, but radii do not reach periphery. For center cuts see Plate 3, no. 4.

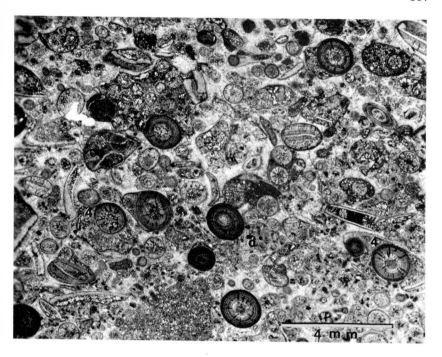

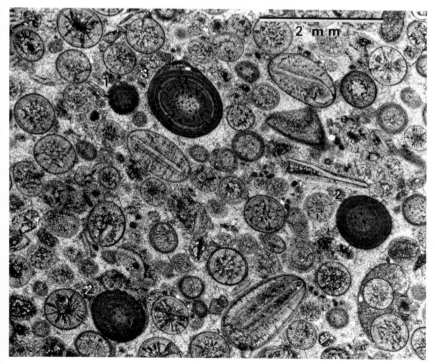

SPHERULITES

Plate 5. Spherulites with radial structures. Bedding horizontal, shows as stylolites, debris, fragmental oöids, and mud (black). Matrix calcite. Aprons grow on oöid surfaces. Deformation was intense enough to overcome randomly oriented primary ellipticity, and a uniform direction is noticeable. AB cut. Ratio 1.69. Ratios of single oöids vary from 1.04 to 1.93, a clear indication of nonspherical oöids before deformation. Sample 85G2, Berryville, Virginia, Area 3.

Plate 6. Section AC normal to Plate 5. Spherulites smaller, structures radial, matrix fairly coarse calcite. Bedding at 60° to oöid elongation barely visible as a zone of dolomite grains, fragmental oöids, and general imperfection. Aprons are very small. Radial structures are growths of calcite, dust particles, and impurities. A few mud pellets of same size as the oöids are also elongated but less regular. Ratio A/C 2.07 for cut. Sample 85G2, Area 3.

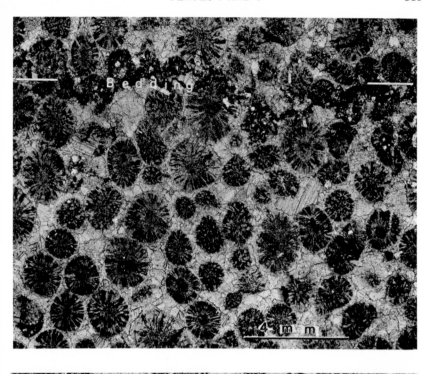

SPHERULITES

Plate 7. Closely packed spherulites with coalescing wedges shown by calcite twinning of larger units at A-ends. Diagonal fractures begin to show. Little interference between oöids. Where spherulites touch there is little solution, stylolitization, interpenetration, or flattening of oöids. Matrix calcite, some mud. Ratio A/C 1.44 for sample, 1.8 for photograph. Sample 812Hb, west of Waynesboro, Virginia, Area 2.

Plate 8. Fibrous growth on black mud pellets and spherulites with centers of different sizes and shapes. Spherulites lack concentric layers, have sharp contacts, radial fibrous growth, and under polarized light a distinct cross. Spherulite cores are crystals, small fragments, or small mud pellets. Average ratio for 100 spherulites is 1.68. One of the two large mud pellets consists of black mud with small carbonate grains, the other is mud with spherulites, mud pellets, and carbonate. Shape of large pebbles has not been changed visibly because they are too large and across elongation direction. Bedding is boundary between dark closely packed oöids with a muddy matrix and widely spaced oöids in a calcite matrix. Large pellets parallel bedding. Matrix is calcite which does not grow in continuity with spherulite growth, but fills space between oöids. Only small pellets well-oriented parallel spherulite orientation. Sample 7226G, northeast of Broadway, Area 2.

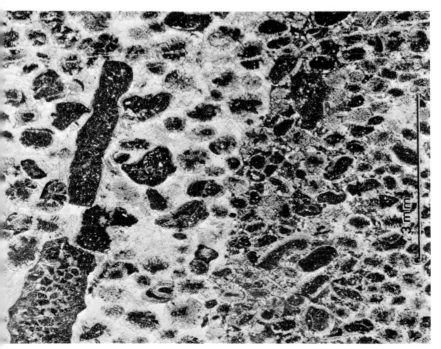

SPHERULITES

Plate 9. Layering on spherulite cores. Centers are radial, contacts between centers and outer layers are sharp. Radial structures: dust rows and fibrous growths that cut layers to periphery. Compare radial structures with Plates 43, 44. Elongated centers are probably fossil fragments with fibrous growths and later layering. Matrix calcite, some mud. Unsharp spherulites are high or low cuts with widely spaced rings and radial structures showing as black dots. Ratio A/C 1.16. Sample 7256Ea, south of Harrisonburg, Virginia, River cut, Area 2.

Plate 10. Spherulitic cores with radial structures and outer layers, mud pellets, and fragments in matrix of mud and calcite. Pellets and fragments oriented parallel to oöids. Ratio A/B about 1.5. Sample 86B4b, Berryville, Virginia, Area 3.

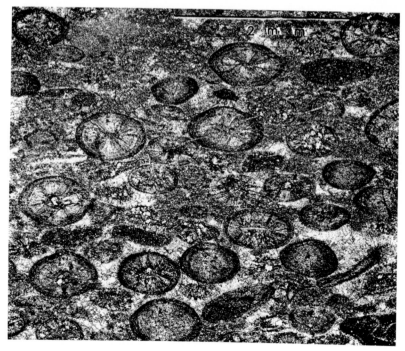

LAYERED OÖIDS

Plate 11. Mixture of layered oöids with spherulitic cores, pellets, and fragments. Some large fragments still show primary nonspherical orientation. Small pellets are elliptical with long diameters parallel to A. Coalescing fractures asymmetrical to A about 57–60 and 15°. Asymmetry suggests counter-clockwise rotation. Matrix calcite. Small aprons on oöids. Ratio A/C 1.56. Sample 7256Fa, south of Harrisonburg, Virginia, Area 2.

Plate 12. Mixture of spherulites with and without mantle, small pellets (black), quartz grains (white), and dolomite. Bedding is a concentration of dark pellets, light quartz, and solution of oöids. Ratio A/C 1.8. Sample 466a, Shepherdstown, east of Martinsburg, West Virginia, Area 3.

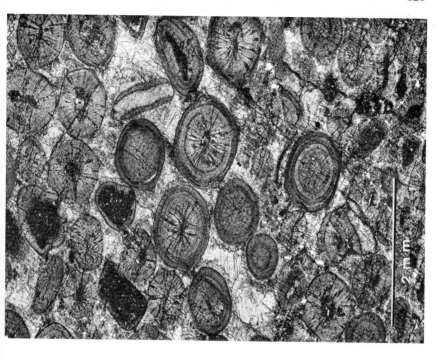

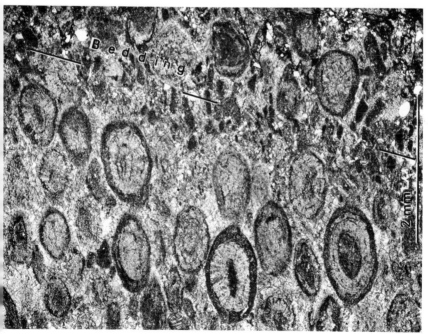

LAYERED OÖIDS

Plate 13. One spherulite in polarized light, slightly deformed within a detached mantle. Center cut. Parallel particles indicate cleavage in oöid mantle and matrix. At the ends mantle begins to fray. Adjacent oöids are high or low cuts showing only mantle with grain orientation parallel to cleavage. Matrix crystalline calcite. Ratio A/C for slide 1.86. In photograph: about 2.4 for mantle, 1.3 for core, naked spherulites 1.81, pellets 4.35. Sample 713Gb, west of Waynesboro, Virginia, Area 3. (See also drawings, Figures 5 and 9.)

Plate 14. Poorly sorted spherulites mostly with mantles, layered oöids, mud pellets, undeformed dolomite, and pebbles. The large oölite pebble (P) shows internal orientations of oöids parallel to outside orientation, but the pebble retained a shape probably parallel to bedding. Matrix particles and oöids share orientation. White crystals are dolomite. Ratio A/C 2.09. Sample 886Cb, Shepherdstown, West Virginia, Area 3.

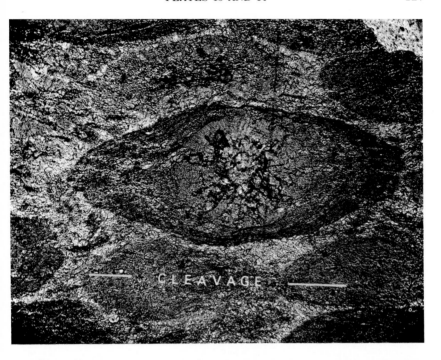

LAYERED OÖIDS

Plate 15. Undeformed irregularly shaped pellets in crystalline calcite matrix. Large slabby pieces consist of small pellets that have not separated. Their orientation suggests bedding. Most pellet surfaces show some growth of a layer of fibrous calcite normal to it. No orientation. Sample 7226E, east of Broadway, Area 2.

Plate 16. Large, slabby mud lumps are roughly parallel, but small fragments are well aligned, normal to large ones. Smallest lumps occur preferably on one side of large ones, as if they either settled on them or broke away. Orientations become more nearly perfect with decreasing lump sizes. Large calcite grains fill voids and are unoriented. No peripheral growth on mud surfaces. Ratio A/C for small pellets 2.23. Sample 6157B, Edinburg, Virginia, Area 2.

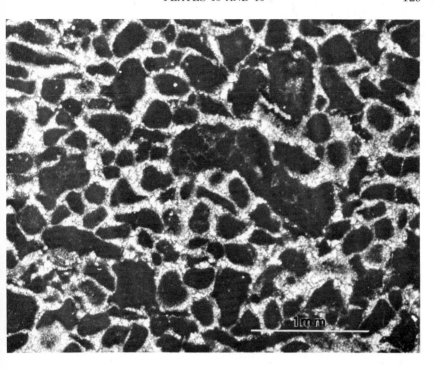

PELLETS

Plate 17. Well-oriented pellets, layered oöids, and spherulites of varying sizes. Odd primary shapes have been overcome. Matrix calcite. No calcite aprons on pellets. Oriented particles within pellets suggest formation of cleavage. Ratio A/C 2.5–3.00. Sample 466Ea, Shepherdstown, West Virginia, Area 3.

Plate 18. Uneven-sized pellets with quartz grains. Small pellets have become cores of spherulites, large ones show fibrous growth rind. Orientation of spherulites uniform with ratio A/C of 1.53. Large pellets show less growth and no orientation. Slabby fossil fragment suggests bedding. Matrix coarse calcite. Sample 7226F, east of Broadway, Area 2.

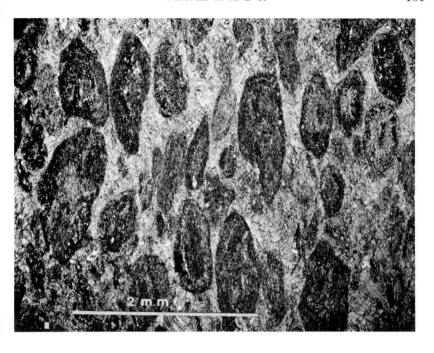

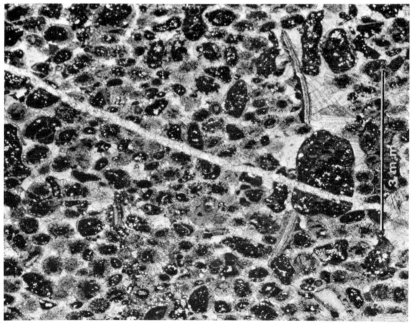

PELLETS

Plate 19. Well-aligned spherulitic oöids with mud centers. Clusters of oöids are "pebbles," and oöids within pebbles are elliptical and parallel with oöids outside. The whole pebble is not elliptical. Very uneven beddomg (white, dashed) is suggested by stylolites, an accumulation of insolubles, and partially dissolved oöids. Small components are well aligned. Matrix is crystalline calcite. Ratio A/C 1.52. Sample 7266M, 5 miles east of Harrisonburg, Area 2.

Plate 20. Small pellets (as centers of spherulites), large unoriented pellets, quartz grains with secondary growths. Ellipticity A/C feeble. Matrix is sparry calcite with some twinning. Quartz overgrowths reach beyond oöids. Ratio A/C 1.38. Sample 7158Cb, south of Broadway, Area 2.

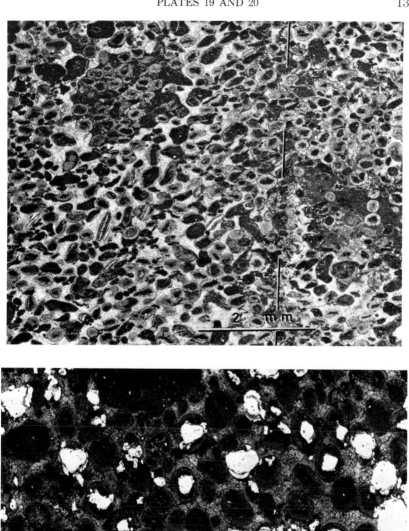

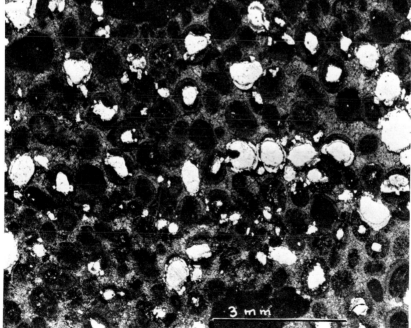

PELLETS AND QUARTZ GRAINS

Plate 21. Dolomite clusters are nearly spherical and covered with a growth rind of dark fibrous calcite. Spherulites are much smaller and well oriented. Fossil fragments and mud pellets randomly oriented, but also with fibrous growth rind. Matrix crystalline calcite. Only spherulites can be used. Ratio A/C for spherulites 1.35. Sample 7226B, east of Broadway, Virginia, Area 2.

Plate 22. Undeformed dolomite clusters within elongated pellets, spherulites, and layered oöids. Dolomite ratio barely measurable, about 1.02 with tendency toward dolomite elongation across pellet elongation. Ratio A/C 2.14. Sample 8166Ea, Lexington-Buena Vista area, Mt. Moriah Church Section, Area 3.

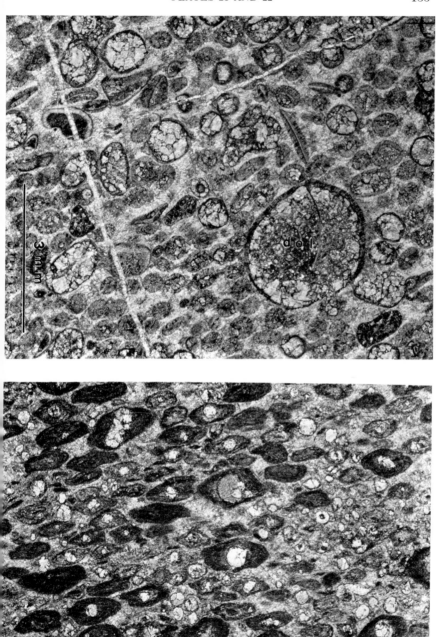

DOLOMITE

Plate 23. Dolomite centers in highly deformed spherulites with layered mantles. Spherulites still show dark cross in polarized light. Mantles have thinned along sides and thickened at the ends. Measurements are at best difficult due to loss of mantle along sides and ends where mud migrated away from spherulite core. Cleavage seen as black lines, parallel elongation of spherulites, mantles, and as particle orientation in mantles and matrix. Spherical dolomite clusters remained undeformed or broke into angular fragments. Ratio for mud mantle A/C 4.57. Sample 687Ba, north of Shenandoah, Virginia, Area 3.

Plate 24. Mantled spherulites, dolomite centers, bedding, and extreme elongation. Frayed ends make measurements difficult and probably minimal. A at least 15° with bedding. Ratio for spherulites about 5.54. Sample 8166Ba, east of Lexington, Virginia, Moriah Church Section, Area 3.

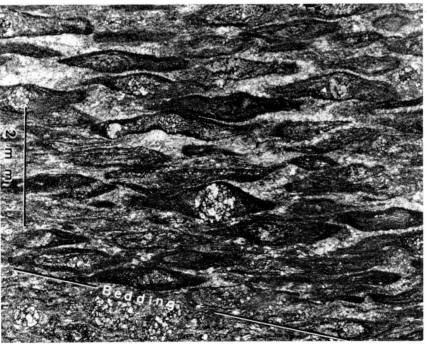

DOLOMITE

Plate 25. Portion of a chert nodule in calcite-mud matrix. Ellipticity of chert oöids is normal to that of pelletal oöids. At contact, mud mantle oöids are preserved as aggregates of carbonate crystals in a mantle of dark, fine-grained mud. The mud oöids become more distorted away from the nodule. Pellets have been dragged by rotation of chert nodule. Centers in pellets are dolomite and calcite spherulites. A/C ratios are chert 1.13, spherulitic calcite 2.25, but increasing with distance from nodule, pellets with centers 4.88. Measurements are difficult and few. Sample 7216B, southwest of Elkton, Virginia, Area 3.

Plate 26. Large chert nodule. Ratio A/C within 1.75 to 2.00. Matrix flows around nodule, consists of leftover calcite and dolomite centers in mud. Within chert nodule long oöid axes are 25° between right and left. Two tension fractures are filled with calcite and widen toward periphery. Rotation is counter-clockwise, similar to that shown in Plate 25. Sample 7865B, Harrisonburg quadrangle.

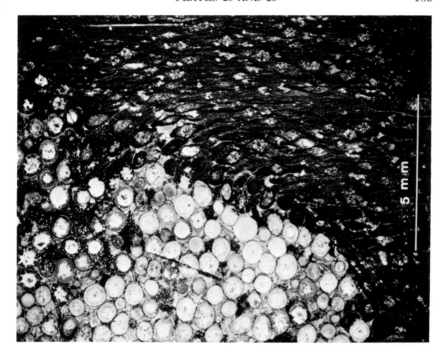

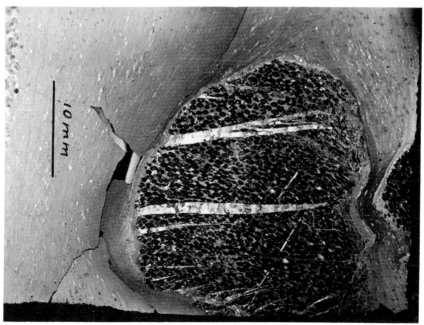

CHERT NODULES

Plate 27. Elliptical mud pebble with detrital sand grains and quartz overgrowth reaching across pebble contact into matrix. Ratios of spherulites and pebble are the same order of magnitude—1.47 and 1.37. Ratio for slide 1.40 with 100 measurements. Sample 7226Ab, south of Mt. Jackson, Virginia, Area 2.

Plate 28. Mud pebbles consisting of oöids, pellets, detrital quartz, and mud in spherulitic oölite with pellets. Matrix calcite. Pebble elongate parallel bedding and normal spherulite A. Sample V65, exact locality lost. Area 2.

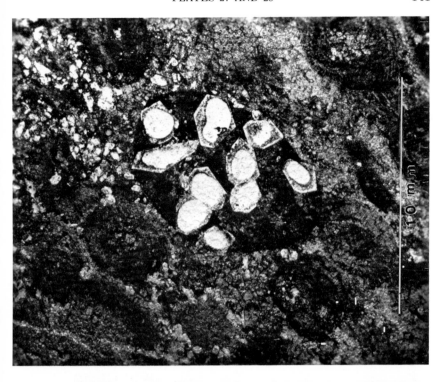

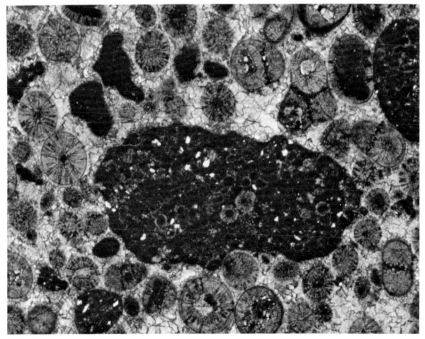

PEBBLES

Plate 29. Large pebble of spherulites in black mud matrix in spherulitic oölite with slabby, unoriented fossil fragments. Matrix calcite. Ratio A/C 1.41 in pebble and outside. Sample 7266Ab, east of Harrisonburg, Virginia, Area 2.

Plate 30. Pebble in spherulitic oölite consists of spherulites, mostly with mud core similar to those outside, but smaller and densely packed. Ratios and orientation equal in pebble and host oölite. Pebble was as ductile as host, deformation did not flow around pebble as with chert nodules and dolomite clusters. The pebble differs from host in spacing of spherulites, number of mud pellets, and granular mud. Pebbles are generally darker than host. A/C ratios, 50 each: mud pellets in pebble 3.08; mud pellets outside 3.15; spherulites in pebble 2.35; spherulites outside 2.27; mud centers in layered oöids 3.09. Sample 813A2, west of Waynesboro, Virginia, Area 3.

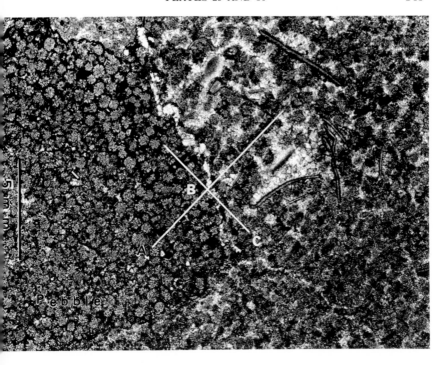

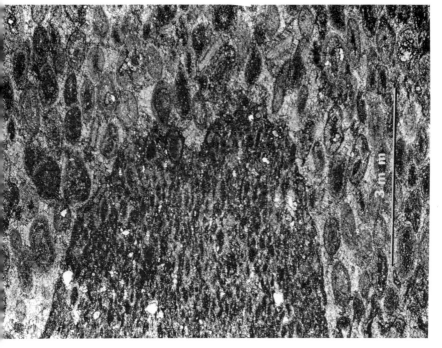

PEBBLES

Plate 31. Large fossil fragment filled with pellets and spherulites in spherulitic oölite with mud pellets, pebbles, assorted debris, stylolites, and calcite vein. A varied collection of components. Ratio A/C for spherulites 1.48. Sample 7266G, northeast of Harrisonburg, Virginia, Area 2.

Plate 32. Distorted fossil in oölite with mud matrix. Spherulites within elliptical fossil cross section show equal ratio as matrix. Cleavage visible as parallel arrangement of all particles. Ratio A/C 3.12. Sample 620C, north of Charles Town, West Virginia, Area 3.

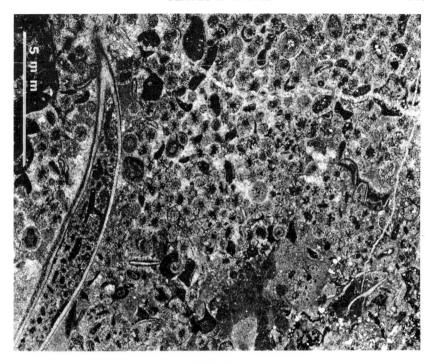

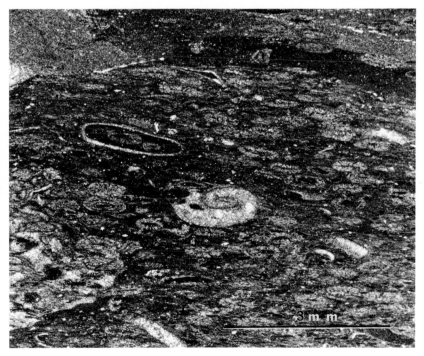

FOSSIL FRAGMENTS

Plate 33. Extremely irregular bedding plane due to post-deformational stylolites. Solution crosses deformed spherulites. Lower layer: mud pebbles and oölite showing cleavage as particle orientation. Elongation parallel in both beds and across stylolite. Ratio A/C 1.71 (slide). Sample 713F, west of Waynesboro, Virginia, Area 3.

Plate 34. Bedding plane separates two distinct lithologies: (1) small-oöid, spherulitic oölite with mud cores, pellets, and layered oöids, and (2) much larger spherulitic oölite with similar components, mud, and layers of dolomite crystals. Elongation normal to bedding. Ratio A/C 2.10 (slide). Sample 886Cb, Sharpsburg, Maryland, east of Winchester, Virginia, Area 3.

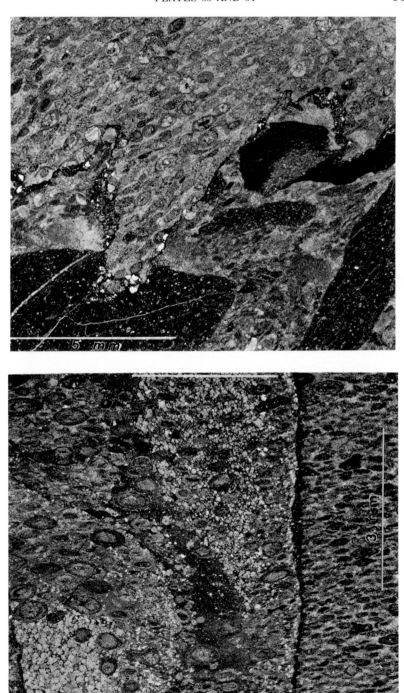

BEDDING

Plate 35. Extremely deformed oölite. Spherulites and mantled oöids can still be recognized in a network of sparry fragments. Dolomite centers are almost intact. Layers of calcite and dark mud are draped over less distorted lenticular cores. Cleavage intense in outcrop. Only well-defined oöids were measured. Ratio A/C 5.84. Sample 717E, southeast of Luray, Virginia, Area 3.

Plate 36. Fairly well defined, very elongated, dark layered oöids with spherulite centers. Fraying ends make measurements difficult, because mantles terminate in streaks. Cleavage intense in outcrops and seen as alignment of particles in photograph. In AC surfaces the rock is not readily recognized as an oölite. Ratio A/C about 5 to 7. Sample 896B, northeast of Charles Town, West Virginia, Area 3.

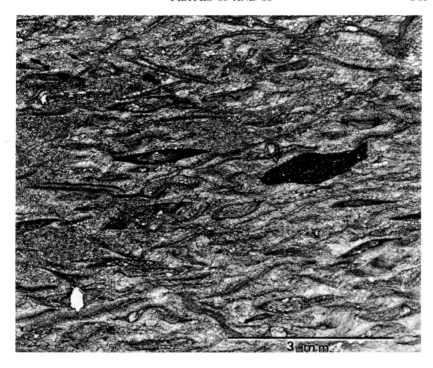

EXTREME DEFORMATION

Plate 37. Spherulites with and without layers and small mud pellets with common orientation. Spherulite center cuts show radial structures due to granular radii between either fibrous or partially crystallized calcite. A cross is seen in polarized light. High or low sections show fracturing or dark areas parallel to A and only feeble radial structures. Boundaries are not sharp, probably due to the gentle dip of surfaces into thin section. Matrix clear, crystalline calcite. A/C ratio for 100 spherulites 1.53; small mud pellets 2.30. Sample 8176Ga.

Plate 38. Oölite with well-defined layered oöids (right half), some radial structures in the cores covered by five or six dark and light layers of carbonate and mud. A clear calcite grain may be in the center. Quartz shows overgrowth partly beyond oöid border with fibrous calcite growing in A. Layered oöids are scattered openly in a light matrix of crystalline calcite with many wispy mud pellets, strongly oriented. The slide appears "dirty." Portion of large pebble fills half of thin section and is filled with small elongated mud pellets which continue orientation in matrix across the pebble. Pebble boundary sharp with stylolites which cut across orientation. A/C ratio of 100 layered oöids 2.91; 50 layered oöids 2.78; 50 mud pellets outside 3.74; 50 pellets inside pebble 3.63. Sample 886G2, Shepherdstown, West Virginia, Area 3.

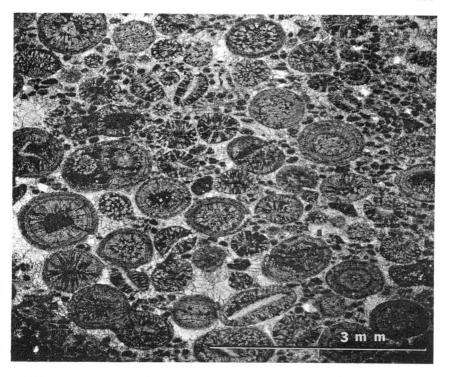

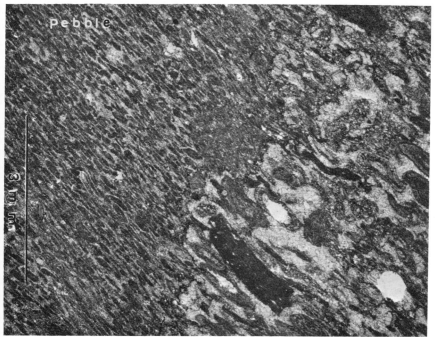

PELLETS, SPHERULITES, AND PEBBLES

Plate 39. A pebble is embedded in a host oölite consisting of well-defined spherulites with distinct boundaries in a crystalline matrix. In the pebble are spherulites of the same size but more closely packed in a dark muddy matrix. The pebble contact is sharp with residue concentrated on stylolites. Contact cuts across spherulites on both sides. Elongation cuts across contact without deflection. On the spherulites calcite makes aprons of crystals. Some aprons grow to twice the diameter of the spherulite, but then the calcite is twinned making herringbone patterns. A/C ratios: inside pebbles 2.19; outside 2.00. Sample 813A4, west of Waynesboro, Virginia on U.S. 340, Area 3.

Plate 40. Sharply defined chert oöids with halos of quartz fibers in an oölite that consists of unassorted debris, long, drawn-out mud pellets with ratios of maybe 10 to 1, quartz grains, crystalline calcite, fossil fragments, and some dolomite. Some chert oöids still show radial rows of inclusions, a center, and halos, but in polarized light they are granular chert without a spherulite cross. Chertification at an early stage has prevented additional deformation. Mud pellets flow around the chert area. Ratio A/C for chert 2.14; for spherulites 2.29; mud pellets beyond measurement. Sample 728A, between Berryville and Front Royal, Virginia.

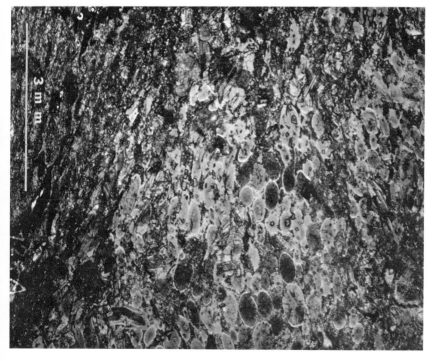

CHERTIFICATION

Plate 41. A layer in an oölite is partially chertified and contains a nodule of completely chertified oöids. Matrix spherulites show a cross in polarized light, centers, and radial structures. Mud lumps include quartz grains and are without shells, growth margins, or well-defined shape. They are derived from odd-shaped fragments. The matrix is clear calcite. Deformation in chert much less than in outside spherulites. A/C ratios: chert, 1.51; remnants in host, 1.74; spherulites, 1.96; mud pellets, 3.76. Sample 5814B, Old Chapel near Berryville, Virginia, Area 3.

Plate 42. Undeformed, almost spherical dolomite clusters with sharp contacts. Normal ratio in that locality is about 2; 100 dolomite clusters 1.22. Sample 8176Ea, east of Lexington, Virginia, Area 3.

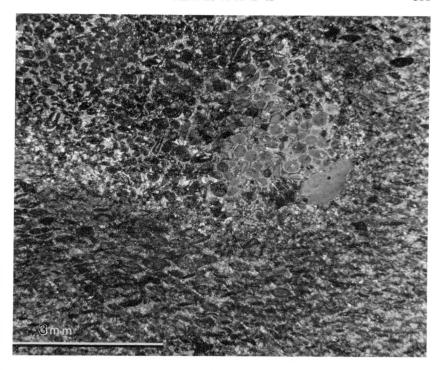

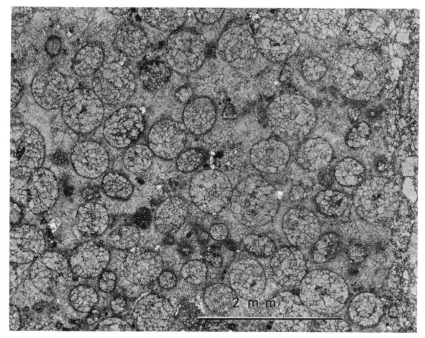

CHERT AND DOLOMITE

Plate 43. Slightly deformed spherulites. Parallel light. Spherulite 1 is laminated; radial structures are like spokes of a wheel. Black material is very fine-grained mud with calcite. Oöid section cut through center because spokes reach from center to periphery, thinning outward. Cut 2 is high or low, and spokes are cut across. Lamination is wider, less distinct, and the oöids are smaller. Elongated oöid 3 is a center cut, and radial structure is determined by shape of center. Sample 7256C, south of Harrisonburg, Virginia, Area 2.

Plate 44. In polarized light the axial cross is pronounced in cut 1; in cut 2 it is feeble and not supported by radial spokes. The elongated spherulite shows a cross, but it is much wider east-west than north-south because of the primary orientation of fibers.

Fractures are barely suggested at the top of the elliptical oöid. They are black lines that reach the periphery. Matrix is not fractured and shows no growth on oöid surfaces. Ratio A/C 1.10. Sample 7256Cb, west side of Massanutten Mountain, south of Harrisonburg, Virginia, Area 2.

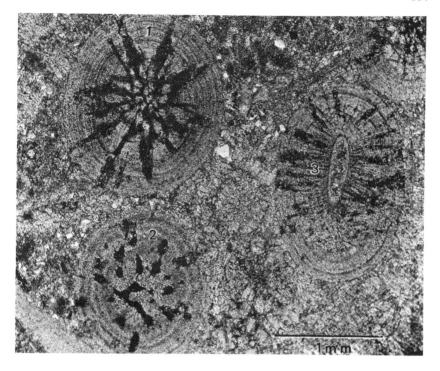

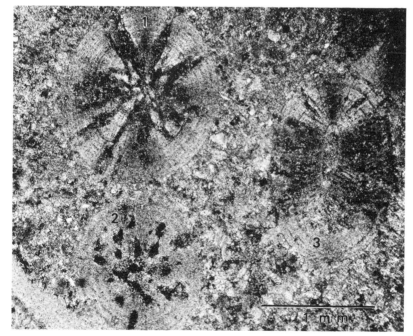

RADIAL STRUCTURES

Plate 45. Several spherulites in parallel and polarized light. Boundaries sharp. Black radial structures reach periphery in center cuts. Between "wheel spokes" fibers consolidated into wedges of one crystal each, as shown by the calcite cleavage which differs between wedges but is uniform within each. Rotation of wedges is minimal, but surfaces are slightly displaced in some oöids. Fractures begin to appear as dark lines, but without uniform orientation.

Plate 46. Under polarized light the axial cross is very distinct and extends into the matrix. A small calcite apron has grown as part of the spherulite. Average diameter of oöids $\frac{1}{2}$ mm. Ratio A/C 1.20. Sample 5288D, northwest of Winchester, Virginia, Area 1.

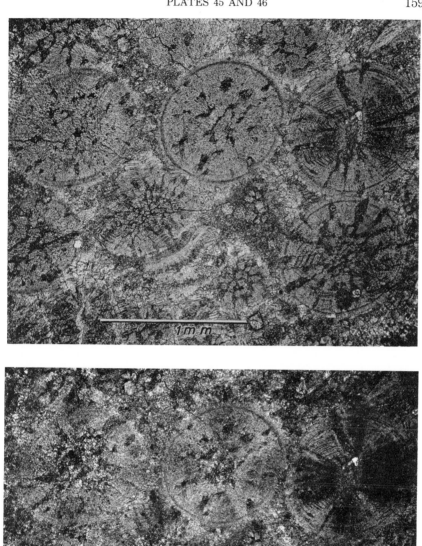

SPHERULITES WITH WEDGES

Plates 47 and 48. One spherulite in parallel light (Pl. 47) and polarized light (Pl. 48). Center cut. Boundary very sharp. Black radial structures reach periphery. Surface unbroken.

In polarized light the cross extends far beyond the periphery, showing that the matrix is partially a growth apron that has grown beyond the spherulite boundary and across the matrix, to the next oöid.

Fractures begin near center of spherulite and continue across matrix to next oöid. A diagonal fracture cuts across spherulite, matrix, and into neighbors. An east-west fracture coalesced across photograph. Other fractures are less continuous. Ratio A/C 1.40. Sample 630D, north of Charles Town, West Virginia, Area 3.

SPHERULITES AND FRACTURES

Plate 49. Spherulites consist of one-crystal wedges. Growth aprons extend into matrix, especially in extension direction, as seen by herringbone pattern and polarized light. Wedges have rotated, and surfaces are faulted as in spherulites Nos. 2, 4, and 6.

Fractures coalesce across matrix as between Nos. 2 and 4. Almost a continuous fracture from No. 1, across 2, and to No. 4. A second diagonal connects No. 4, across No. 6 to 7. Spherulites clearly facilitate fracturing across their centers. No rotation visible. Sample 420 (1947). Ratio A/C 1.31. Spring Mills, W. Va., Area 2.

PLATE 49   163

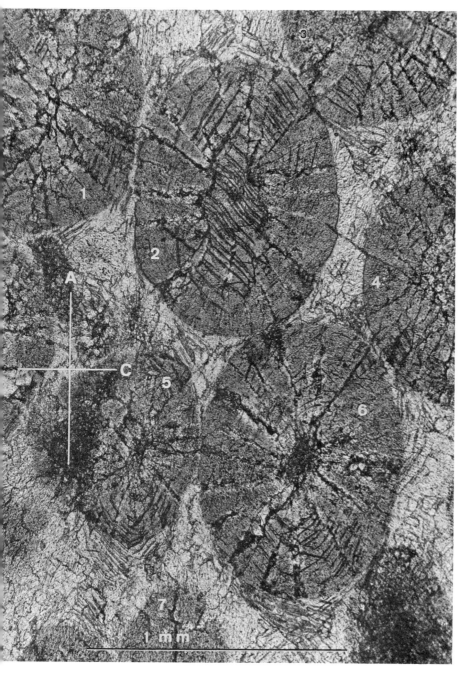

FRACTURE PATTERNS

Plate 50. A/C ratio has increased to 2.37. Radial fractures were rotated at the ends, compressing the wedges between them as in a vise. The center became elongated. Particles in the groundmass are parallel to A. At the ends of the spherulites new carbonate grains have grown on 4 of the 5 pictured oöids.

Two kinds of fractures occur: one derived from radii limited to the oöids and a second one cutting across the entire slide. The radial fractures at the ends displace the oöid surfaces. Wedges have been rotated and tilted and the dark area moved away from the center.

Two carbonate veins occur: a tension joint normal to A filled with secondary calcite and open; and a narrower vein in the position of a left shear fracture about 23° from A. The two fractures may be shears at 62°: the wide one at high angle (antithetic), the other one at low angle (synthetic) in a counter-clockwise rotation accompanying simple shear. Sample 371a (1947), east of Hagerstown, Maryland, Area 3.

Plate 51. Oöids are much distorted. The center spherulite is elongated and fractured, its center is a long, black gash. Radial fractures bound wedge-shaped crystals with uniform calcite cleavage. The surface is displaced on radial fractures. In the central area wedges have been rotated to the right above center and to the left below it. This could indicate clockwise rotation. Light, radiating tension fractures with secondary carbonate formed before deformation terminated. Sequence is as follows: (1) deformation and rotation of oöid wedges with displacement of surface; (2) tension joints normal to A; and (3) additional rotation of all parts, including the tension joints.

The angles between early radial fractures and A are 18 and 22°, the angle between white tension gashes and A 60 to 70°. The tension gashes formed normal to A and were rotated about 20 to 30°, the early radial fractures may have rotated 60 or 70°. The crystal wedges have been tilted like dominoes, and this became possible due to the extension of A, since dominoes standing upright take less space than tilted ones. Other examples of white tension gashes: Plates 54, 55. A/C ratio 3.38. Sample 344 (1947), south of Shippensburg, Pennsylvania, Area 3.

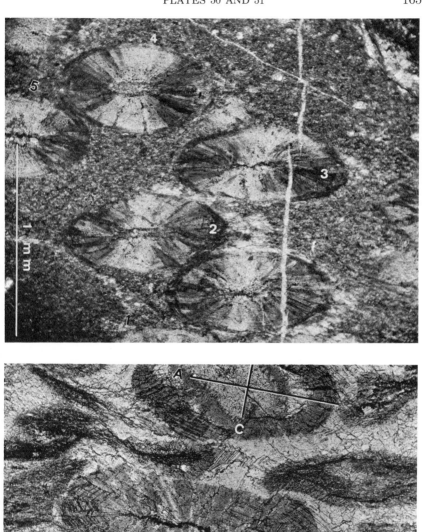

DEFORMATION MECHANICS

Plate 52. AB section, Plate 53 AC section of the same sample. Sections are normal to each other, 2 cm apart. A/C ratio 1.56.

Plate 52, primary radial structures but no fractures. Oöids are laminated without growth aprons. Plate 53 (AC) shows distinct fractures in oöids and matrix. Oöid 3 is a center cut, radial fractures reach the surface but do not displace it. Oöid 2 is cut off center but near it. Fractures are 20 and 70° from A, and the same directions are suggested in oöids 1 and 4, which are cut higher or lower than 2. Similar directions are suggested in the matrix.

Comparison of Plates 52 and 53 shows that the fractures of Plate 53 are not visible in the section normal to Plate 53. The fractures intersect in B but are not prominent enough to show in a AB section. They are visible only in "profile." Sample 7256Fb, 8 miles south of Harrisonburg, Virginia, Area 2.

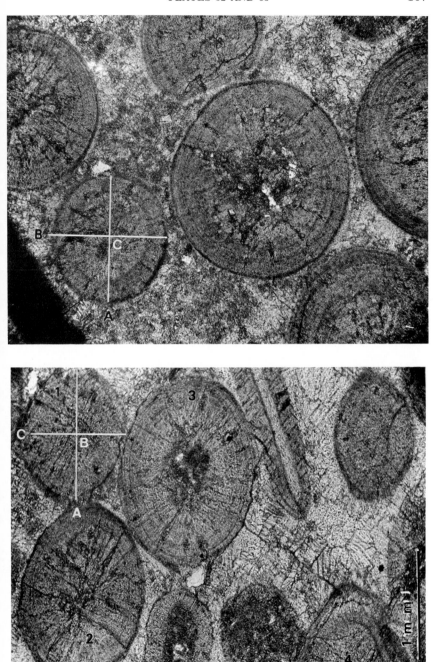

FRACTURES AND SECTIONS

Plate 54. AC cut of Plate 51. Wedges with individual cleavage have rotated toward *A*, and the oöid surface is faulted. The oöid has been flattened. A large apron has grown on the oöid, as shown by calcite cleavage. Light lines cut across *A* and suggest tension fractures with new white calcite. Large stylolite with black insolubles and debris cuts across orientation. Parallel particles very distinct in lower left corner of photograph. Black lines parallel to long axis are cleavage.

Plate 55. AB cut, same sample. Ratio is less. Fractures cut across center, wedges are slightly displaced, but twinned overgrowth at A-ends is as prominent as in AC section. The white tension fractures are wider and more pronounced than in AC cut. The center has been stretched parallel to *A*, the two ends were compressed between rotating fractures. Ratios: A/C 3.38; A/B 1.60. Sample 344, south of Shippensburg, Pa., Area 3.

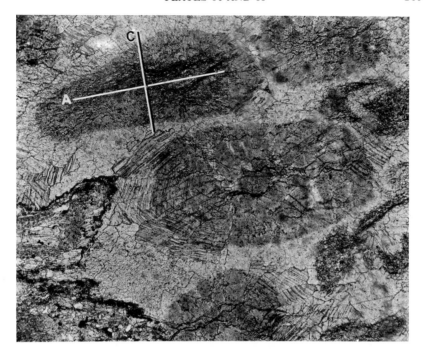

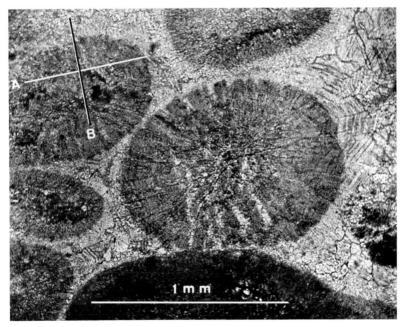

FRACTURED SPHERULITES

Plates 56 and 57. Slightly deformed oöids in off-center sections with very feeble orientation. Plate 56, radial fractures like faults above salt domes. Fibrous structure very distinct. Plate 57, polarized light. Aprons have grown well into the matrix and are continuous with spherulites' extinction. Large replacement crystals are dolomite. Sample 5288, northwest of Winchester, Virginia, Area 2.

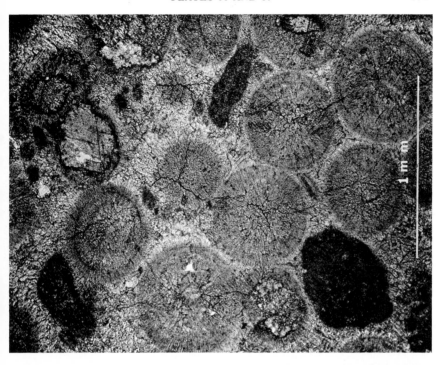

FRACTURED SPHERULITES

Plate 58. Fracture pattern across slide symmetrical to elliptical spheru-lites. Fractures coalescing into shears penetrate fossil fragment at right. Ellipse No. 1 high or low cut just begins to show diagonal fractures. El-lipse No. 2 faulted with wedge moving toward center. Ellipse 3 untrust-worthy because of elongated core. Fragments are randomly oriented in slide, no suggestion of bedding. Angle between fractures about 70°. A/C ratio 1.34. Sample 58–16, Winchester area, Area 2.

Plate 59. Diagonal fractures distinct in spherulite 1. High cuts 2 and 3 also diagonally fractured. Fractures emphasized by granular debris and mud, or zones of small particles. Radial fractures outline wedges. Particle orientation near 4 suggests cleavage parallel to long axes. Locality un-known.

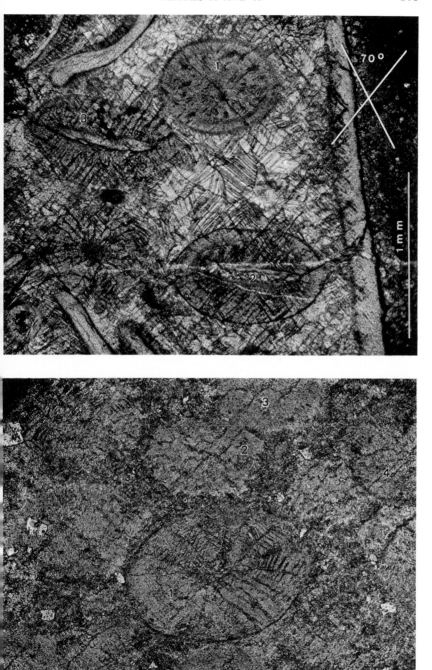

COALESCING FRACTURES

Plate 60. Calcite fibers grow on quartz grain (black) parallel to cleavage in intensely deformed, lineated limestone. (*Note*: one dark spherulitic, elliptical oöid below quartz grain, suggesting a ratio A/C more than 3.) Sample 58–6, Limeton, Va., Area 3.

Plate 61. Quartz sand layers in schistose limestone. Fold hinge at right with axial plane cleavage. Calcite grows on both ends of quartz grains. Right quartz-grain layer is folded. Cleavage is axial plane cleavage in that fold. (*Note*: difference of elongated black pellets, quartz shards with calcite growth, and undeformed quartz.) Beyond measurement. Sample 8166A. East of Lexington, Va., Area 3.

MINERAL GROWTH AND CLEAVAGE

Plate 62. Link between slickensided cleavage plane and elliptical oöids. Streaks and patches are parallel to long ellipsoid axes of inserts. At T: tension fractures normal to striae. Insert 1 is a photomicrograph of sample. Oöid with center about 2 mm long. Insert 2 almost the same scale as sample. Stria are also direction of a co-ordinate, and AB plane is ab in the area. A/C ratio 5–6. Sample 717C, south of Luray, Virginia, road to Antioch church, Area 3.

PLATE 62   177

LINK BETWEEN ÖOIDS AND SLICKENSIDES

Plate 63. Surface of specimen of "gneiss" from Harpers Ferry, W. Va., several hundred feet west of intersection of U.S. 340 and Va 671. Small opening in intensely schistose zone about 50 to 70 feet thick. Above and below, the gneiss is massive with some banding but little schistosity. Black long streaks derived from chloritized garnets. Long axes of streaks dip 20° in 117 (azimuth).

Plate 64. Surface of block is schistosity with dark patches as in Plate 63. Two cuts parallel and normal to lineation are planes of thin sections of Plates 65 and 66. White irregular zones on surface are tension fractures normal lineation.

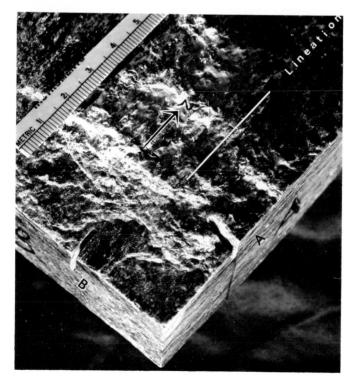

LINEATION IN GNEISS

Plate 65. Photomicrograph of thin section parallel to lineation, normal schistosity. Streakiness that looks like schistosity is mainly lineation. Black wisps are longitudinal section of lineation streaks of Plate 63.

Plate 66. Photomicorgraph of cut B (Pl. 64), normal lineation, Schistosity not distinct, grains isometric, no clusters or streaks. Shows structure in Plate 65 to be linear.

PHOTOMICROGRAPHS OF GNEISS

Plate 67. Aligned, elliptical chlorite blebs in greenstone. South Mountain, Maryland. A:B:C approximately: 30:4:.1 mm.

Plate 68. Sericite patches in tuffaceous Catoctin beds. Note small well-aligned particles in matrix. East of Elkton, at entrance to National Park. Swift Run Gap road.

LINEATION IN VOLCANICS

Plate 69. Even-sized sericite patches in tuffaceous beds. Ellipsoid ratios approximately 20:10:.1 mm. Catoctin Formation, east of Elkton, Virginia.

Plate 70. Uneven-sized small chlorite blebs in Catoctin Greenstone and parallel mineral growth in cleavage plane grading into slickenside-like mineral growth in matrix between blebs. Lower portion of sample is striated. Scale as in Plate 69.

LINEATION IN VOLCANICS

Plate 71. Large, drawn-out, black chlorite patches in cleavage surface. White lines normal to lineation are crenulations and tension fractures. Surface suggests transition to slickenside and not well-defined distorted bodies.

Plate 72. Surface of slabby Loudoun conglomerate with elongated pebbles, small sericite blebs in pebbles, and lineation in matrix. South Mountain, Maryland, east of Blue Ridge Summit on Pa. 16, several hundred feet downhill from Western Maryland Railroad crossing.

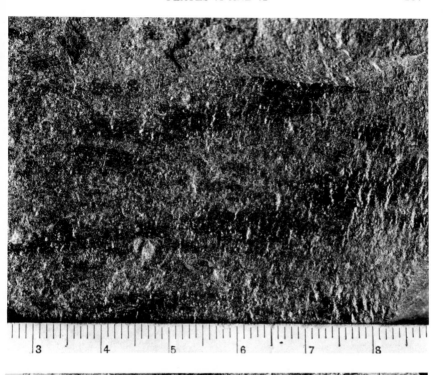

Plate 73. Cleavage surface in Loudoun Formation with lineation normal to bedding trace. Lineation points 120° within cleavage which dips 20°. Bedding dips steeply northwest. South Mountain, east of Blue Ridge Summit, Pa., Road 16 from Summit to Emmitsburg, Maryland.

Plate 74. Lineated cleavage plane in Conococheague Limestone. About 50 feet above schistose limestone are deformed oölites with ratio A/C up to 4.15. A is parallel lineation. Sample 716A, Ingham section, Area 3.

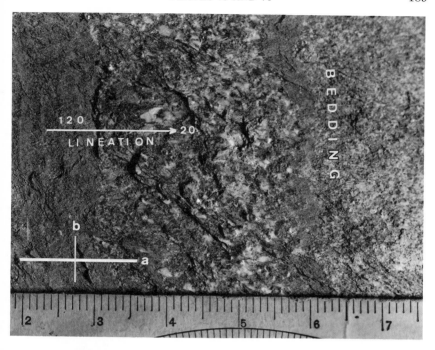

LINEATIONS

Plate 75. Lineation in Conococheague Limestone dips 62° west. Intersection of U.S. 340 and U.S. 11, opposite Howard Johnson Restaurant. Sample 813E, south of Staunton, Virginia, Area 3.

Plate 76. Bulging cleavage in Catoctin Greenstone. Mile 55, Skyline Drive. Scale: Prof. Eugene Wegmann, Neuchâtel, Switzerland.

LINEATION AND BULGING CLEAVAGE

Plate 77. Striated surface in Weverton Formation. Surface is scratched and coated with secondary mineralization. Mineral orientation continues in adjacent sandstone.

Plate 78. Slickensided joint in Athens Limestone at Alma, Virginia, west of bridge over Shenandoah River. A thick coating of calcite is slickensided, accompanied by growth of calcite at several levels. Coating about 2 cm thick. Scale: cm.

SLICKENSIDES

Plate 79. Fractured lens of epidosite with actinolite and quartz fibers growing in fractures. Fractures are parallel to *b* and normal to *ac*. Actinolite grows parallel to front surface *ac*. Scale: cm, see also Fig. 20.

Plate 80. Two fractures intersect in co-ordinate *b* with rounded corner. Striations in both surfaces are growth rods of quartz in up to 2-cm-thick patches. Rods grow in intersection of the fractures with *ac* surface. Mile 55, Skyline Drive, Shenandoah National Park, Virginia.

Plate 81. Greenstone with quartz rods protruding on rough surfaces. Tension joints normal to rods. Two cuts were made: *A* is parallel with rods and scale (Pl. 82); *B* is normal to them. Scale: cm.

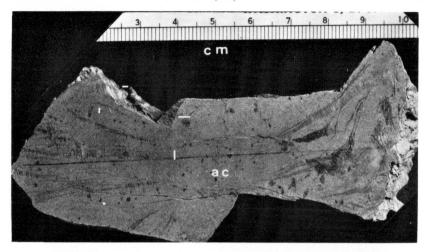

GROWTH IN FRACTURES

Plate 82. Photomicrographs of a thin section cut parallel to rods and scale of Plate 81. A fracture opened, and as quartz rods grew the top moved to the left, rotating rods into an inclined or almost parallel position. The quartz vein is about 2 cm thick, and the rods broke off on tension joints in the sample surface. Sample 7197E, Rapidan Fire Road, Big Meadows, Shenandoah National Park.

Plate 83. Fracture in greenstone was filled by about 8 or 10 quartz crystals. Subsequently the large grains were broken into fibers, and actinolite grew parallel with them. In addition fibers were folded along flexures photomicrograph. Sample 7187, Rapidan Fire Road, Big Meadows, Shenandoah National Park.

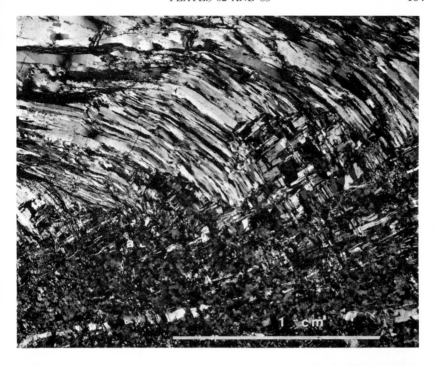

QUARTZ IN FRACTURES

Plate 84. Catoctin Formation unstrained flow top amygdules filled with quartz. Ellipticity of amygdule is primary-prefilling. Big Meadows, Shenandoah National Park.

Plate 85. Elliptical amygdules suffered additional strain which broke quartz into needles. Same scale for Plates 84 and 85. Big Meadows, Shenandoah National Park.

QUARTZ IN AMYGDULES

Plate 86. Amygdule fill disintegrating into needles which radiate from one center. For scale, see Plate 87.

Plate 87. Quartz rods in greenstone. Large grains are disintegrating into needles or breccia. Sample 8227C, Mile 61, Skyline Drive.

STRAINED QUARTZ

Bibliography

Albrecht, Karl, 1966. Ungleichartiges Verhalten kompetenter und inkompetenter Schichten bei tektonischer Gesteinsumformung: *Geol. Mitteilungen* 6: 213–38.

Albrecht, Karl, and Furtak, Hans, 1965. Die tektonische Verformung der Fossilien in der Faltenmolasse Oberbayerns zwischen Ammer und Leitzach: *Geol. Mitteilungen* 5: 227–48.

Allen, Rhesa M., Jr., 1967. Geology and Mineral Resources of Page County, Virginia: Virginia Division of Mineral Resources, Charlottesville, Virginia, Bull. 81.

Ampferer, O., 1906. Über das Bewegungsbild von Faltengebirgen: *J. Österr. Geol. Reichsanstalt* 56: 539.

Balk, Robert, 1952. Fabric of Quartzites near Thrust Faults: *Jour. Geol.* 60: 415–35.

Benz, K. C., 1932. Über Chamosit und Thüringit von Schmiedefeld, Thüringen, und ihre genetischen Beziehungen: *Jahrb. Mitt. Oberrhein., Geol. Ver., Stuttgart*, N. F. vol. 12, xii: 20–31.

Berg, G., 1944. Vergleichende Petrographie oolitischer Eisenerze: *Arch. Lagerstättenforschg. H.* 76, Berlin, Reichsamt f. Bodenforschung.

Bick, Kenneth F., 1960. Geology of the Lexington Quadrangle, Virginia: Virginia Division of Mineral Resources, Charlottesville, Virginia, Rept. of Investigations 1.

Billings, Marland P., 1954. *Structural Geology*. New York: Prentice-Hall.

Breddin, Hans, 1931. Über das Wesen der Druckschieferung im Rheinischen Schiefergebirge: *Centralblatt f. Min.*, Abt. B, pp. 202-16.

———, 1956. Die tektonische Deformation der Fossilien im Rheinischen Schiefergebirge: *Zeitschr. Deutsche Geol., Ges. v-106:* 227–305.

———, 1957. Tektonische Gesteinsverformung im Gebiet von St. Goarshausen (Reinisches Schiefergebirge): *Decheniana* 110: 289–350.

———, 1962. Petrogene Mineralgänge im Paläozoikum der Nordeifel und ihre Beziehungen zur inneren Deformation der Gesteine: *Geol. Mitteilungen* 2: 197–224.

———, 1964. Die tektonische Deformation der Fossilien und Gesteine in der Mollasse von St. Gallen (Schweiz): *Geol. Mitteilungen* 4: 1–68.

———, 1967. Quantitative Tektonik I. Einführung, II, Allgemeine tektonische Verformungen: *Geol. Mitteilungen* 7: 205–38.

———, 1968. Quantitative Tektonik 2. III. Faltung: *Geol. Mitteilungen* 7: 333–436.

Breddin, Hans; Furtak, Hans; and Hellerman, Eberhard, 1964. Eine geometrische Erklärung für die flache Lage der Faltenachsenflächen und der Schiefrigkeit: *Geol. Mitteilungen* 3: 253–74.

Brent, William B., 1960. Geology and Mineral Resources of Rockingham County, Va.: Virginia Division of Mineral Resources, Charlottesville, Virginia, Bull. 76.

Bryan, W. H., and Jones, O. A., 1955. Radiolaria as Critical Indicators of Deformation: Univ. Queensland Papers, vol. IV, no. 9.

Bühl, G. 1956. Lagerstätten-Kundliche Untersuchungen des Eisenerzvorkommens von Schmiedefeld/Thüringen: Unpubl. Dipl. Arb. Everberg 1959.

Carter, Neville L., and Friedman, Melvin, 1965. Dynamic Analysis of Deformed Quartz and Calcite from the Dry Creek Ridge Anticline, Montana: *Am. Jour. Sci.* 263: 747–85.

Christie, John M., and Raleigh, C. B., 1959. The Origin of Deformation Lamellae in Quartz: *Am. Jour. Sci.* 257: 385–407.

Cloos, Ernst, 1940. Crustal Shortening and Axial Divergence in the Appalachians of Southeastern Pennsylvania and Maryland: *Geol. Soc. America Bull.* 51: 845–72.

———, 1946. *Lineation, a Critical Review and Annotated Bibliography:* Geol. Soc. America Mem. 18.

———, 1947. Oölite Deformation in the South Mountain Fold, Maryland: *Geol. Soc. America Bull.* 58: 843–918.

———, 1950. The Geology of the South Mountain Anticlinorium, Maryland: Guidebook 1, Johns Hopkins Univ. Studies in Geology, no. 16, pt. 1, Johns Hopkins Press.

———, 1951. Structural Geology of Washington County, in *The Physical Features of Washington County,* pp. 124–63. Maryland Department of Geology, Mines and Water Resources.

———, 1955. Experimental Analysis of Fracture Patterns: *Geol. Soc. America Bull.* 66: 241–56.

———, 1957. Blue Ridge Tectonics between Harrisburg, Pennsylvania and Asheville, North Carolina: *Proc. Natl. Acad. Sciences* 43: 834–39.

———, 1964. Appalachenprofil 1964. *Geol. Rundschau* 54: 812–34.

———, 1966. Dating Blue Ridge Deformation Plan with Slickensides and Lineations (Abstract): Geol. Soc. America, Philadelphia, Pennsylvania meeting.

———, 1968. Experimental Analysis of Gulf Coast Fracture Patterns: *Am. Assoc. Petrol. Geol. Bull.* 52: 420–44.

Cloos, Ernst, and Hershey, H. Garland, 1936. Structural Age Determination of Piedmont Intrusives in Maryland: *Proc. Natl. Acad. Sci.* 22: 71–80.

Cloos, Ernst, and Hietanen, Anna, 1941. Geology of the "Martic Overthrust" and the Glenarm Series in Pennsylvania and Maryland: Geol. Soc. America, Spec. Paper 35.

Cooper, Byron N., 1964. Relation of Stratigraphy to Structure in the Southern Appalachians: in *Tectonics of the Southern Appalachians:* V.P.I., Dept. Geol. Sciences Mem. 1, pp. 81–114.

Dean, S. L., 1966. Geology of the Great Valley of West Virginia: Diss. W.Va. Univ., Morgantown, W.Va.

Dennis, J. G., 1967. International Tectonic Dictionary: Am. Assoc. Petrol. Geol. Mem. 7.

Dziedzie, K., 1964. The Geological Significance of the Orientation of Pebbles: *Geol. Sudetica* 1: 301–7.

Ellenberg, Jürgen, 1964. Beziehungen zwischen Oöid-Deformation in den ordovizischen Eisenerzen und der Tektonik an der SE-Flanke des Schwarzburger Sattels (Thüringen): *Geologie,* Jahrg. 13: 168–97.

Elliott, David, 1970. Determination of Finite Strain and Initial Shape from Deformed Elliptical Objects: *Geol. Soc. America Bull.* 81: 2221–36.

Engels, B., 1956. Zum Problem der tektonischen Verformung der Fossilien im Rheinischen Schiefergebirge: *Deutsche Geol. Ges. Z.* 106: 306–7.

Fellows, Robert E., 1943. Recrystallization and Flowage in Appalachian Quartzite: *Geol. Soc. America Bull.* 54: 1399–1432.

Fisher, George W.; Pettijohn, F. J.; Reed, J. C., Jr.; Weaver, Kenneth N., editors, 1970. *Studies of Appalachian Geology, Central and Southern:* New York: Interscience Publ., John Wiley.

Flinn, D., 1956. On the Deformation of the Funzie Conglomerate, Fetlar, Shetland: *Jour. Geology* 64: 480–505.

———, 1961. On Deformation at Thrust Planes in Shetland and the Jotunheim Area of Norway: *Geol. Mag.* 98: 245–56.

Furtak, Hans, 1966. Der Aufschluss in den oberdevonischen Famenneschichten unterhalb der St.-Adelbert-Kirche (Kaiserplatz) in Aachen: *Geol. Mitteilungen* 6: 37–42.

Furtak, Hans, and Hellermann, Eberhard, 1961. Die tektonische Verformung von pflanzlichen Fossilien des Karbons: *Geol. Mitteilungen* 2: 49–69.

Graf, D. L., and Lamar, J. E., 1950. Petrology of Fredonia Oölite in Southern Illinois: *Am. Assoc. Petrol. Geol. Bull.* 34: 2318–36.

Helmers, J. H., 1955. Krinoidenstielglieder als Indikatoren der Gesteinsdeformation: *Geol. Rundschau* 44: 87–92.

Henderson, John R., 1969. Tectonic Significance of Minor Structures in the Conococheague Formation near Lancaster, Pennsylvania. *Am. Jour. Sci.* 267: 166–81.

Hills, E. Sherbon, 1963. *Elements of Structural Geology:* New York: John Wiley.

Hubbert, M. K., and Rubey, W. W., 1959. Role of Fluid Pressure in Mechanics of Overthrust Faulting: *Geol. Soc. America Bull.* 70: 115–66.

King, P. B., 1950. Tectonic framework of southeastern United States: *Am. Assoc. Petrol. Geol. Bull.* 34: 635–71.

————, 1950. Geology of the Elkton area, Virginia: U.S. Geol. Survey Prof. Paper 230.

————, 1959. *The Evolution of North America.* Princeton: Princeton Univ. Press.

————, 1964. Further Thoughts on Tectonic Framework of Southeastern United States, in *Tectonics of the Southern Appalachians:* V.P.I. Mem. 1, pp. 5–31.

Kurtman, F., 1960. Fossildeformation und Tektonik im nördlichen Rheinischen Schiefergebirge: *Geol. Rundschau* 49: 439–59.

Kvale, Anders, 1957. Gefügestudien im Gotthardmassiv und den angrenzenden Gebieten (vorläufige Mitteilung): *Schweiz. Min. Petr. Mitt.* 37: 398–434.

————, 1966. Gefügestudien im Gotthardmassiv und den angrenzenden Gebieten: Schweiz. Geotechnische und Geol. Kommission.

Lake, P., 1943. Restoration of Original Form of Distorted Species: *Geol. Mag.* 80: 139–47.

Lee, F. T., 1961. Geology of the Front Royal area, Warren County, Virginia: MS thesis, V. P. I., Blacksburg, Va.

Lindström, Maurits, 1955. A Tectonic Study of Mt. Nuolja, Swedish Lapland: *Geol. Fören. Förhandl.* 77: 557–66.

————, 1955. Structural Geology of a Small Area in the Caledonides of Arctic Sweden: Inst. of Mineral., Paleont., Geology, Univ. Lund, Nr 27.

————, 1957. Tectonics of the Area between Mt. Keron and Lake Allesjaure in the Caledonides of Swedish Lapland: Inst. of Mineral., Paleont., and Quat. Geology, Univ. Lund, no. 37.

————, 1958. Tectonic Transports in the Caledonides of Northern Scandinavia East and South of the Romrbak-Sjangeli Window: Inst. of Mineral., Paleont., and Quat. Geology, Univ. Lund, no. 43.

Loretz, H., 1885. Erläuterungen zur Geol. Specialkarte von Preussen und den Thüringischen Staaten: Lfg. XXX, Bl. Spechtsbrunn, Berlin.

Nickelsen, R. P., 1956. Geology of the Blue Ridge near Harpers Ferry, West Virginia: *Geol. Soc. America Bull.* 67: 239–70.

Perkins, R. L., 1967. Tectonites from the Blue Ridge and Great Valley of West Virginia and Northwestern Virginia: Ph.D. thesis, West Va. Univ., Morgantown, W.Va.

Pray, L. C., and Murray, R. C., editors. 1965. Dolomitization and Limestone Diagenesis, a Symposium: Soc. Econ. Paleont. and Mineral., Spec. Publ. no. 13.

Ramsay, D. M., 1964. Deformation of pebbles in Lower Old Red Sandstone Conglomerates adjacent to the Highland Boundary Fault: *Geol. Mag.* 101: 228–48.

Ramsay, John G., 1967. *Folding and Fracturing of Rocks:* New York: McGraw-Hill.

Reed, John C., Jr., 1955. Catoctin Formation near Luray, Virginia: *Geol. Soc. America Bull.* 66: 871–96.

————, 1969. Ancient Lavas in Shenandoah National Park near Luray, Virginia: U.S. Geol. Survey Bull. 1265.

————, 1970. The Blue Ridge and the Reading Prong, Introduction: *in* Fischer, Pettijohn,

Reed, and Weaver, *Studies of Appalachian Geology, Central and Southern*, pp. 195–99. New York: Interscience Publ., John Wiley.

Rodgers, John, 1949. Evolution of Thought on Structure of Middle and Southern Appalachians: *Am. Assoc. Petrol. Geol. Bull.* 33: 1643–54.

———, 1953. The Folds and Faults of the Appalachian Valley and Ridge Province, *in* McGrain, Preston, *et al.*, Ky. Geol. Survey Spec. Publ. 1, pp. 150–66.

———, 1964. Basement and No-basement Hypotheses in the Jura and the Appalachian Valley and Ridge, in *Tectonics of the Southern Appalachians*, V.P.I. Mem. 1, W.D. Lowry, editor, pp. 71–81.

———, 1967. Chronology of Tectonic Movements in the Appalachian Region of Eastern North America: *Am. Jour. Sci.* 265: 408–27.

———, 1970. *The Tectonics of the Appalachians:* Regional Geology Series, New York, Interscience Publ., John Wiley.

Rogers, W. B., 1884. A reprint of annual reports and other papers on the geology of the Virginias: New York.

Sander, Bruno, 1930. *Gefügekunde der Gesteine*. Berlin: Jul. Springer.

Stauffer, Mel R., 1970. Deformation Textures in Tectonites: *Canadian Jour. of Earth Sciences* 7: 498–511.

Stevens, G. R., 1959. Nature and Distribution of S-Planes in Maryland and Pennsylvania: Ph.D. thesis, The Johns Hopkins Univ.

Turner, F. J., 1957. Lineation, Symmetry, and Internal Movement in Monoclinic Tectonite Fabrics: *Geol. Soc. America Bull.* 68: 1–18.

Wellman, H. W., 1962. A Graphical Method for Analyzing Fossil Distortion Caused by Tectonic Deformation: *Geol. Mag.* 99: 348–52.

Werner, H. J., 1966. Geology of the Vesuvius Quadrangle, Virginia: Virginia Division of Mineral Resources, Charlottesville, Va., Bull. 7.

Whitten, E. H. Timothy, 1966. *Structural Geology of Folded Rocks*. Chicago: Rand McNally.

Wickham, John S., 1969. Structural Geology of the Western Slope of the Blue Ridge near Front Royal, Virginia: Ph.D. thesis, The Johns Hopkins Univ.

Wilson, G., 1951. The Tectonics of the Tintagel Area, North Cornwall: *Quart. Jour. Geol. Soc.* 106: 393–432.

———, 1961. The Tectonic Significance of Small Scale Structures and their Importance to the Geologist in the Field: *Ann. Soc. Geol. Belg.* 84: 423–548.

Wunderlich, H. G., 1964. Mass, Ablauf und Ursachen orogener Einengung am Beispiel des Rheinischen Schiefergebirges, Ruhrkarbons und Harzes: *Geol. Rundschau* 54: 861–82.

———, 1965. Mass, Ablauf und Ursachen orogener Einengung am Beispiel der Westalpen und des Nordapennin: *Geol. Rundschau* 55: 699–715.

Appendix: Raw Data

(Abbreviations: sp: spherulites; lsp: layered spherulites; m: mud pellets; dol: dolomite; ch: chert ooids; *: pictured samples)

Sample	Cut	Cut surfaces				Thin sections				Abbreviations + Plates
		# Meas.	Long axis	Short axis	Ratio	# Meas.	Long axis	Short axis	Ratio	
58-1	AC	30	20.57	11.87	1.73	35	16.9	10.7	1.57	sp
	AB	30	21.43	14.97	1.43	30	24.83	19.17	1.30	
58-2	AC					27	18.1	5.11	3.54	sp
	AC					30	26.60	6.03	4.41	
58-3	AC	30	14.70	7.17	2.05	poor slide				sp
	AB	30	13.83	10.30	1.34					sp
	AC	30	14.70	6.83	2.15					sp
58-3a	AB					30	18.33	13.30	1.38	
	AB					40	47.05	33.50	1.41	
58-3b	AC					30	20.20	9.30	2.17	sp
	AC					40	52.18	20.95	2.49	
58-6	*AC	30	20.33	14.73	1.38				>3.00	Pl. 60
58-10	AC	30	19.33	8.97	2.16	40	68.02	18.47	3.68	m
58-12a	AC	30	9.80	6.27	1.56	21	28.09	5.95	4.72	
58-12b	AC	30	10.30	5.70	1.80	30	22.43	9.57	2.34	
58-12c	AB					raw surface				
58-12d	AC					30	23.90	8.43	2.83	sp dol
-13a	AC									
-13b	AC					30	12.30	6.30	1.95	dol
58-14a	AB					30	20.13	13.73	1.47	
-14b	*AC					50	17.96	9.16	1.96	sp
58-14b	BC					40	12.20	7.00	1.74	ch Pl. 41
	AC					20	13.05	8.6	1.51	ch
58-14b	AC					100	15.87	11.26	1.41	lsp, sp
	AC					100	19.77	9.88	2.00	lsp, sp
58-14d	AC					100	19.95	11.75	1.70	lsp, sp
	AC					30	20.37	11.86	1.72	lsp, sp

208

Sample		n				n				Notes
58-15	AC					30	13.37	11.27	1.19	sp
58-16	*AC					30	17.87	13.37	1.34	sp Pl. 58
58-16	AC					30	16.90	12.50	1.35	sp
58-17	AC					30	12.33	10.23	1.21	lsp
58-18	AC	30	14.17	11.93	1.19	30	16.57	13.27	1.25	sp
58-19	AC	30	17.47	14.83	1.18					lsp
58-20a	AC	30	9.60	7.40	1.29					lsp
-20c	AC	30	15.53	12.97	1.19	30	15.97	13.73	1.16	sp
58-21	AC	30	13.43	11.40	1.18	30	20.40	15.33	1.33	lsp
58-22	AC	30	9.90	8.50	1.16					
58-23cc	AC	30	13.97	11.27	1.24	30	15.63	13.43	1.16	lsp
58-24	AC					100	22.03	15.19	1.45	sp
58-25	AC	30	21.90	6.20	3.53	30	25.40	13.27	1.91	m
58-26	AC					30	12.93	4.23	3.05	lsp
58-27	AC		mud, directions		5.11	30	20.83	12.40	1.68	
58-28a	AC	30	23.23	4.37	5.32	30	17.13	6.73	2.54	m, dir only
58-28b	AC	30	20.20	12.57	1.61	30	29.87	8.97	3.33	lsp
58-29	AC					40	39.87	13.75	2.90	m, poor
58-32a	AC					100	27.98	8.70	3.21	sp
58-32b	BC					100	23.68	8.52	2.77	
58-41	AC	100	34.48	7.84	4.39	30	18.70	11.33	1.65	lsp
58-42a	AC	200	19.09	7.10	2.75	40	46.82	26.30	1.78	lsp
	AB	100	19.50	13.50	1.44	30	24.03	10.00	2.40	m, dir
	BC	100	14.10	8.75	1.90	20	37.85	25.45	1.49	lsp
58-42b						20	37.55	23.15	1.62	lsp
-43a	AC	100	22.20	13.40	1.65	100	23.41	14.47	1.61	
-43b	BC					30	22.70	16.90	1.34	lsp

Appendix table—*Continued*

Sample	Cut	# Meas.	Long axis	Short axis	Ratio	# Meas.	Long axis	Short axis	Ratio	Abbreviations + Plates
58-44	AC	100	14.72	5.00	2.94	100	21.03	15.03	1.39	
	AC						20.74	16.03	1.29	
41363-1	BC					30	19.23	8.30	2.32	m
	AC					30	18.76	6.06	3.09	lsp, dol
	BC					30	16.20	5.23	3.09	
41363-2	AC					30	17.13	12.60	1.36	sp
	BC					30	15.50	11.93	1.29	
41363-3	AC					30	19.43	14.10	1.38	sp
	AB					30	18.86	14.36	1.31	
	BC					30	15.13	13.13	1.15	
62065-B	AC	30	16.40	4.53	3.62					lsp
62065-C	*AC					30	16.86	5.40	3.12	sp, m Pl. 32
629-A	AC	100	11.30	2.33	4.84					m?
63065-C	AC	30	8.50	5.27	1.61					sp, m
	BC	30	7.67	4.80	1.59					
63065-D	AC					30	26.43	18.10	1.46	sp
	*AC					100	28.39	20.27	1.40	sp Pls. 47, 48
63065-E	AC					30	26.83	19.43	1.38	sp
	AC					100	27.51	20.44	1.34	sp
63065-F	AC					30	31.77	15.13	2.09	lsp
7165-A	AC	30	16.60	6.30	2.63					?
	AC					30	23.70	5.73	4.13	sp, m
7165-B	AC					60	27.10	6.00	4.51	sp

Sample	Form	n			ratio	n			ratio	habit	plate
7165-C	AC	30	21.63	18.27	1.18	30	16.90	11.73	1.44	sp	
7165-D	AB	30	17.17	12.07	1.42	30	23.83	19.87	1.19	sp	
7165-E	AC	100	23.86	17.14	1.39					lsp	
7165-F	AC	30	19.97	16.17	1.23					lsp	
7165-I	AC	30	18.63	14.00	1.33	30	14.80	8.90	1.66	sp	
7865-A	AC	30	7.00	3.03	2.31	30	22.57	10.73	2.10	lsp, sp	
	BC	30	10.87	6.53	1.66					m	
7865-B	AC					30	25.26	9	2.80	ch	Pl. 26
	*AC					50	23.42	13	1.80	sp	
7865-C	AB	30	28.27	11.5	2.46	30	22.50	12.43	1.81	m	
7865-D	BC	30	12.30	3.67	3.35					lsp	
713-A	AC	100	18.29	8.47	2.16	30	14.47	9.07	1.59	lsp	
713-B	BC	100	18.50	12.04	1.53	30	22.13	13.77	1.61	lsp	
	AB	30	20.77	7.61	2.71	30	19.03	7.23	2.63	lsp	
	AC	30	17.10	11.10	1.54					lsp	
	BC	30	13.80	7.57	1.82					lsp	
713-C	AC	30	22.47	5.73	3.92	100	18.22	7.67	2.37	lsp	
	AC	100	22.38	7.63	2.93					lsp	
	AB	30	23.37	12.13	1.92					lsp	
713-D	AC					30	17.33	12.33	1.4	sp	
	AC					100	15.28	11.30	1.35	sp	
	BC					30	15.93	11.77	1.35	?	
713-E	AC	100	18.21	9.95	1.83	30	19.13	14.90	1.28	lsp	
713-F	*AC	30	17.80	10.27	1.73	100	16.50	9.64	1.71	lsp	Pl. 33

Appendix table—Continued

Sample	Cut	# Meas.	Long axis	Short axis	Ratio	# Meas.	Long axis	Short axis	Ratio	Abbreviations + Plates
713-G	AB					30	43.70	27.67	1.57	lsp
	AC					30	42.97	15.93	2.69	lsp
	AC					100	40.11	14.92	2.68	lsp, sp & m
	*AC					100	20.52	10.99	1.86	lsp, sp in m Pl. 13
	AC					69	19.44	9.97	1.95	lsp, sp
713-H	AC	30	26.17	3.90	6.71					m
	AB	30	34.30	9.53	3.59					m
713-I	AC	100	19.31	8.41	2.29	30	15.93	11.53	1.38	lsp
713-J	AC	30	18.87	11.5	1.64					lsp
713-K	AC									lsp, dol
714-A	AC					30	13.97	7.50	1.86	lsp
	AB					30	11.17	8.30	1.34	
714-B	AB	30	13.93	5.20	2.68					lsp, m
	AB	100	18.14	10.88	1.66					lsp, m
	AC	30	27.17	5.40	5.03	30	32.07	7.07	4.54	lsp, m
	AC	100	18.76	4.34	4.32					lsp, m
714-D	AC	30	13.33	6.66	2.00					sp?
715-A	AC	30	18.2	7.9	2.30	30	18.7	10.03	1.86	lsp
715-B	AC	30	16.80	9.23	1.82					lsp
	AB	30	17.5	14.3	1.22					
716-A	AC	30	15.90	3.87	4.11	30	16.00	10.10	1.58	lsp
	AB									
716-B	AC	30	17.27	9.8	1.76	30	18.66	6.87	2.72	lsp
	AB					30	20.83	12.27	1.70	
716-C	AB	30	25.33	14.46	1.75					lsp
	AC	30	28.13	6.7	4.2					lsp

212

Sample		n							
716-D	AB	30	32.8	15.73	2.08				lsp
	AC	30	27.86	7.53	3.7				lsp
716-F	AB	30	17.93	10.16	1.76				
	AC	30	22.66	5.43	4.17				
717-B	AC	30	36.17	4.5	8.04	39.96	6.13	6.52	lsp
	AC	100	34.45	5.17	6.66				lsp
	AB	30	39.77	16.57	2.40				lsp
717-C	AB	30	38.36	15.83	2.42	39.03	19.40	2.01	sp
	AC	30	27.70	5.43	5.1	30.8	10.1	3.05	sp Pl. 62
	*AC	100	31.08	5.48	5.67	30.90	8.3	3.72	
	BC					17.80	8.13	2.19	
717-E	*AC	30	29.20	5.0	5.84				sp, def v intense Pl. 35
	AB	30	21.96	9.5	2.30				
723-B	AC	100	20.10	3.45	5.82				lsp, m
728-A	*AC	30	16.97	4.13	4.10				m
		100				15.13	7.07	2.14	ch Pl. 40
		26				16.61	6.73	2.47	sp
		30				14.5	6.33	2.29	sp
728-B	AC	30	15.03	6.73	2.23	18.47	8.97	2.06	calculated
	BC	30	11.83	7.23	1.63				
	AC/BC		15.03	7.00	2.16				
728-C	AC	30	17.13	8.00	2.14				lsp, sp
	BC	30	13.63	7.10	1.92				
728-D	AB	30	20.46	13.2	1.55				
	AC	30	19.80	11.6	1.78				
728-E	AC	30	32.00	8.10	3.95	43.26	9.96	4.34	sp
	AC	100				31.77	9.23	3.44	lsp
	AC	35				28.05	8.05	3.48	
	AC	35				39.92	10.58	3.77	m, mantle
728-F	AB	30	29.30	15.2	1.93	27.10	9.33	2.90	sp
	AC	30	20.40	7.43	2.74				
	BC	30	14.13	7.63	1.85				

Appendix table—*Continued*

Sample	Cut	# Meas.	Long axis	Short axis	Ratio	# Meas.	Long axis	Short axis	Ratio	Abbreviations + Plates
728-G	AC	30	19.80	6.26	3.16	100	34.35	6.47	5.31	sp
	BC	30	12.33	5.36	2.3					
728-H	AC	30	29.56	5.73	5.16						m, dol
	AB	30	36.76	16.76	2.21					
728-J	AC	30	14.96	9.53	1.57						sp
	BC	30	13.53	9.9	1.37					
728-L	AC	30	24.10	9.76	2.47					
	AC	100	25.39	9.88	2.56						sp
	AB	30	24.0	15.56	1.54					
728-M	AC	30	16.33	8.50	1.92						sp
	AB	30	15.03	10.73	1.40					
728-O	AC	30	15.40	9.40	1.64						sp
	BC	30	11.40	7.03	1.63					
728-P	AC					40	18.6	5.44	3.34	m
85-B	AC	30	21.76	5.86	3.71						lsp
	BC	30	12.76	5.50	2.32					
	AC	100	12.18	2.26	5.39	11	30.70	6.45	4.76	lsp
	BC	100	6.39	2.04	3.13					
	BC	30	17.36	6.90	2.52					
85-C	AC	100	20.96	10.54	1.99	30	35.30	12.90	2.74	sp
85-E	AB					30	30.1	21.63	1.39	lsp
85-F	AB					30	37.36	23.93	1.56	
	BC					30	26.43	14.50	1.82	
	AC						37.36	13.12	2.84	lsp, calculated
85-F$_2$	AC	30	31.23	9.40	3.32	30	30.83	9.73	3.17	lsp
	BC	30	19.0	9.7	1.9					

Sample		n				n				
85-G	AC					30	22.50	8.73	2.57	lsp
	AC					100	19.53	7.17	2.72	lsp
	BC					30	16.2	8.83	1.84	lsp
	BC					100	15.10	7.90	1.91	lsp
85-G$_2$	AB	10	19.8	14.7	1.35					calculated Pl. 5
	*AC	10	19.8	9.1	2.17	30	15.27	8.87	1.72	
	BC	30	14.4	9.1	1.58					
	BC	100	14.80	8.87	1.69					
86-A$_3$	*AC	30	11.41	5.50	2.07	30	19.03	8.86	2.15	lsp Pl. 6
	AC	30	17.16	6.60	2.60					lsp, dol
	AB	30	20.00	11.23	1.78	100				sp
86-A$_4$	AC	30	18.03	11.06	1.63		21.86	13.61	1.60	
	AB	30	17.93	12.93	1.38		20.95	15.04	1.39	
	AB						20.59	12.09	1.70	
	AB					30	17.03	11.47	1.48	lsp
86-B$_1$	AC	30	34.77	11.30	3.08					
	AB	30	35.50	20.2	1.75					
86-B$_2$	AB	30	24.63	13.70	1.8					
	BC	30	16.63	7.0	2.37					
	BC	100	15.38	6.39	2.40					
86-B$_4$	AC	100	24.28	4.48	5.41	30	46.43	10.77	4.31	Pl. 10
	*AB	30	26.7	10.87	2.46	100	36.48	24.50	1.49	lsp
	AC		30.13	20.00	1.51					
	BC		32.80	10.81	3.03					
86-B$_5$	AC	30	27.80	11.43	2.46	100	25.41	13.99	1.81	calculated
	AC	100	31.36	11.97	2.62	30	21.93	12.33	1.78	lsp
	AB	30	24.8	19.27	1.28		36.48	13.99	2.60	lsp
	AB	100	32.15	21.37	1.50					lsp

Appendix table—*Continued*

Sample	Cut	# Meas.	Long axis	Short axis	Ratio	# Meas.	Long axis	Short axis	Ratio	Abbreviations + Plates
812-A	AB	30	19.33	15.70	1.23					
	BC	30	16.50	10.2	1.62					
	AC	100	18.95	13.27	1.42					
812-B	AC	30	13.00	6.17	2.10					dol
	AC	100	13.30	5.87	2.26					lsp
	AC	22	6.87	3.09	2.22					lsp
	AB	100	14.08	10.34	1.36					lsp
812-C	AC	30	11.77	8.00	1.47	100	12.24	7.95	1.53	sp
	AC	100	7.90	5.09	1.55	100	13.13	10.03	1.30	sp
	AB	30	12.3	7.37	1.66					sp
	BC	100	8.86	5.45	1.62					sp
812-D	AC	30	16.57	11.23	1.47					sp
	AC	100	17.07	10.66	1.60					sp
	AB	100	17.14	14.09	1.21					sp
812-E$_3$	AC	30	14.80	11.03	1.34					sp
	AC	30	13.6	8.10	1.67					sp
	AC	100	13.80	7.15	1.93					sp
	AB	100	13.98	11.60	1.20					sp
	BC	30	11.76	7.97	1.48					sp
	AC	30	15.01	9.80	1.54	30	17.70	10.90	1.62	sp
	AB	30	15.17	12.67	1.19	30	17.53	14.70	1.19	sp
812-F	AB	100	14.57	11.28	1.29	30	13.10	7.67	1.71	sp
	AB	100	13.10	8.12	1.61	100	20.56	14.01	1.46	sp
	AC	100	13.93	8.84	1.57					sp
	AC	100	17.01	11.50	1.48					sp
812-G	AC	30	13.60	9.63	1.41					sp
	AC	30	14.50	8.97	1.62					sp

Sample	Axis	n	L	W	L/W	n	L	W	L/W	Remarks
812-H	AC	100	11.22	7.04	1.59					sp
	BC	30	14.30	9.63	1.48					sp
	BC	30	13.40	8.27	1.62					sp
	BC	100	10.16	6.41	1.58					sp
812-I$_1$	*AC	30	13.00	9.23	1.41	30	22.27	15.50	1.44	sp Pl. 7
	BC	30	10.27	8.07	1.27	30	18.67	13.53	1.38	sp
812-I$_2$	AC	30	13.03	8.23	1.58	30	12.90	8.27	1.56	sp, dol
	AC	100	9.73	6.07	1.60					sp
	AB	100	12.54	6.60	1.90					sp
	AC	100	11.72	6.65	1.76					sp
812-I$_3$	AC	30	9.90	4.87	2.03	30	10.94	6.00	1.82	sp
812-J$_1$	AB	30	13.08	8.31	1.57					sp
	BC	100	8.13	5.38	1.51					sp
	AC		13.08	5.38	2.43					sp
812-J$_2$	AB	30	13.11	9.80	1.34					sp
	AB	100	13.33	10.98	1.21					sp
	BC	100	11.62	7.59	1.53					sp
812-K	AC	30	13.33	7.59	1.75					sp
	AC	30	12.00	5.74	2.08					sp
812-L	AC	30	11.43	5.63	2.03					lsp
	AC	30	16.34	7.72	2.15					lsp
	AC		13.03	6.63	1.96					
812-N$_2$	AB					30	24.76	16.66	1.47	sp
	AC					30	35.00	16.17	2.16	sp
	AC					30	94.00	42.50	2.21	sp
	AC					30	12.33	6.30	1.96	sp
813-A$_1$	AC	30	15.46	8.07	1.92					
	AC	100	12.51	6.66	1.87					
	AB	100	14.46	10.86	1.33					
813-A$_2$	*AC					43	21.67	7.97	2.71	lsp, m
	AC					50	14.86	3.76	3.08	

Appendix table—*Continued*

Sample	Cut	# Meas.	Long axis	Short axis	Ratio	# Meas.	Long axis	Short axis	Ratio	Abbreviations + Plates
813-A$_2$	AC					50	30.88	9.80	3.15	lsp, m Pl. 30
	AC					50	42.04	18.44	2.27	lsp, m
	AC					35	14.57	6.20	2.35	lsp, m
	AC					50	7.92	2.24	3.09	lsp, m
813-A$_3$	AB	30	12.67	7.53	1.60					sp
	AB	100	14.06	10.71	1.31					sp
	AC	100	15.25	8.30	1.83					sp
813-A$_4$	*AC	30	15.10	6.76	2.20	64	16.55	8.26	2.00	lsp Pl. 39
	AC					30	19.50	8.83	2.20	
	AC					65	17.24	7.87	2.19	
813-B	AC					30	44.63	20.93	2.19	sp
813-C	AC	100	16.34	12.47	1.31	30	19.86	13.30	1.49	sp
	BC	100	16.84	13.68	1.23	30	19.86	13.30	1.49	sp
1966										
466-A	AB					30	19.00	5.85	3.24	lsp
	*AC					30	49.84	27.37	1.80	Pl. 12
466-B	AC	100	18.04	9.26	1.94					lsp
	AB	100	18.93	14.16	1.33					lsp
466-C	AC	30	18.23	8.06	2.26	30	29.35	11.53	2.50	lsp
	AC	30	12.23	9.90	1.23					
	AC	30	14.57	9.17	1.59					
466-D	AC	100	16.70	10.03	1.66	30	30.87	21.73	1.42	lsp
	AB	100	18.17	13.67	1.33					lsp
	BC	50	14.58	10.36	1.40					lsp
466-E	*AC					30	35.43	11.56	3.00	sp & m Pl. 17
7206-A	AC	30	12.76	5.36	2.38	37	13.70	6.94	1.97	lsp, m, fossil debr
	AC	100	12.76	5.19	2.45					lsp, m, fossil debr

Sample	Orientation	n				n				Remarks
7206-B	BC	30	15.20	10.43	1.46	100	15.18	8.20	1.85	lsp, m, fossil debr
	AC	30	17.20	6.17	2.79	100	18.46	7.60	2.42	lsp, m, fossil debr
7216-A	AB	30	17.50	12.87	1.36					sp, m
	AC	30	16.03	7.06	2.27	100	16.80	7.92	2.12	
7216-B*	AC	40	46.40	9.50	4.88					ch, m Pl. 25
	BC	40	23.30	20.50	1.13					
7226-A	AB	30	16.00	13.37	1.19	64	16.37	13.48	1.21	sp Pl. 27
	*AC	30	15.90	12.90	1.23	100	20.68	16.83	1.22	
7226-B	*AC	30	10.90	7.76	1.40	100	21.50	15.29	1.40	sp, dol Pl. 21
7226-E	no orientation					100	12.99	9.60	1.35	Pl. 15
7226-F	*AC	30	10.93	6.60	1.66	100	14.12	9.20	1.53	sp, m Pl. 18
7226-G	*AC	30	11.30	7.07	1.60	100	19.64	8.59	2.28	sp, m Pl. 8
	AC					100	14.21	8.42	1.68	sp
	AB					100	16.76	8.36	2.00	m
	AB					100	13.21	9.63	1.37	sp
7226-H	BC	30	12.60	8.96	1.40	100	23.35	19.41	1.20	sp, m, dol
7256-A	AC	30	19.17	16.23	1.18	100	20.48	16.60	1.23	dol
7256-B	AC	30	18.53	15.37	1.21	100	11.78	8.08	1.45	lsp, ch
7256-C$_1$	*AC	30	13.70	9.87	1.39	100	21.61	18.22	1.18	sp
	*AB	30	18.27	14.37	1.27	100	20.67	17.47	1.18	Pls. 3, 4, 43, 44
7256-C$_2$	calculated from AB + AC					33	36.36	30.76	1.18	large sp
	calculated from AB + AC					167	18.13	15.29	1.18	small sp
7256-C$_2$	*AC	30	27.27	20.93	1.30					sp
	AC	30	27.63	20.67	1.34					sp
7256-C$_3$	AB					160	22.07	16.61	1.32	large sp
	AC + AB					118	22.70	18.83	1.20	large sp
	AC + AB					105	34.82	27.72	1.25	small sp
7256-D$_1$	AC	30	28.40	19.33	1.47	173	27.29	21.09	1.29	large sp
	AC	30	13.10	9.03	1.45					small sp

Appendix table—*Continued*

	Cut	# Meas.	Long axis	Short axis	Ratio	# Meas.	Long axis	Short axis	Ratio	Abbreviations + Plates
7256-D₁	BC	30	27.27	18.90	1.44					large sp
	BC	30	13.37	10.43	1.28					small sp
	AB	30	23.60	20.10	1.17					large sp
	AB	30	12.80	10.43	1.23					small sp
7256-D₃	AC	30	30.36	19.20	1.58					large sp
	AC	30	15.00	9.03	1.66					small sp
	AB	30	29.73	22.70	1.31					large sp
	AB	30	16.10	12.47	1.29					small sp
	BC	22	24.14	18.32	1.32					large sp
	BC	30	13.70	10.83	1.26					small sp
7256-E	*AC	30	11.37	9.73	1.17	100	20.82	17.84	1.16	lsp Pl. 9
	AC					100	20.61	16.78	1.22	lsp
7256-F	*AC	30	13.90	9.47	1.47	100	23.56	15.04	1.56	lsp Pl. 11
	*AB					102	25.05	20.97	1.19	lsp Pls. 52, 53
7256-G₁	AC	30	12.57	10.43	1.20	100	14.96	12.98	1.15	dol, sp
	AB					65	15.60	13.21	1.18	
7256-H₁	AC	100	14.42	11.59	1.24	100	17.32	12.94	1.33	sp
	AB									
7256-H₂	AC	30	14.57	11.30	1.29	100	16.26	12.99	1.25	sp
	AB	30	15.53	11.73	1.32	100	14.72	10.37	1.41	sp
7266-A	*AC	30	11.80	8.37	1.41	100	14.18	11.79	1.20	sp Pl. 29
	AB									
7266-B	AC	30	9.17	6.63	1.38	100	10.40	8.06	1.29	sp
	AB	30	8.90	7.23	1.23	100	11.42	9.10	1.25	sp
7266-C	AC	30	13.83	11.53	1.19		13.25	10.93	1.21	sp
	BC	30	12.33	10.33	1.19					sp
	AB	30	12.67	11.67	1.13	100	16.81	13.91	1.21	sp

Specimen	Pos.	N				N				Notes
7266-D	AB	30	19.97	16.67	1.20	100	22.84	19.23	1.18	sp
7266-G	*AC	30	15.57	10.57	1.47	100				sp, m
	AC	100	16.79	11.28	1.48					sp, m Pl. 31
7266-H	AB	100	15.04	12.44	1.21	100	13.88	10.60	1.30	sp, m
	AC	30	13.83	9.63	1.43	100	16.23	11.93	1.36	sp
	AC	30	13.10	10.20	1.28					sp
7266-J	AC					100	11.97	9.15	1.30	sp
	AB					100	11.44	8.59	1.33	sp
	AC					100	11.71	9.00	1.30	sp
7266-K	AC	30	8.47	5.83	1.45					sp
7266-L	*AC	30	15.83	9.40	1.68					
7266-M	AC small					100	15.45	10.13	1.52	sp & m Pl. 19
	AC large					50	12.40	5.01	2.43	
	AB					50	21.74	11.58	1.84	
7266-M	AC	30	12.10	9.83	1.23	100	15.72	12.10	1.29	sp
7266-N	AC	30	11.40	9.06	1.25	100	12.76	10.85	1.17	sp
7266-P	AC	30	26.50	12.00	2.21	100	30.79	12.51	2.46	sp in mud
886-A	AB	30	21.70	17.13	1.27					
	BC	30	20.97	14.33	1.46					
886-C	AB	30	29.77	21.80	1.37	100	31.83	23.99	1.32	lsp Pls. 14, 34
	*AC	30	25.23	12.60	2.00	100	32.76	15.67	2.09	
	BC					100	26.82	16.79	1.59	
886-D	AC	30	19.37	10.10	1.92	100	30.88	18.62	1.65	lsp
	AB	30	27.42	17.13	1.60	100	30.47	19.40	1.57	lsp
	AC	100	15.94	7.08	2.25					lsp
886-E	AC	30	17.60	9.60	1.83	99	15.85	11.14	1.42	sp
	AC					100	21.37	12.23	1.74	lsp
886-F	AB	30	17.13	13.93	1.23	100	19.93	16.17	1.23	lsp
	BC	100	11.80	8.52	1.38					sp
	AB	100	12.58	11.07	1.13					sp
	AC	100	12.68	9.01	1.40					sp

Appendix table—*Continued*

Sample	Cut	# Meas.	Long axis	Short axis	Ratio	# Meas.	Long axis	Short axis	Ratio	Abbreviations + Plates
886-G$_2$*	AC					50	25.48	9.14	2.78	lsp, m Pl. 38
	AC					50	11.76	3.14	3.74	lsp, m
	AB					50	20.44	5.62	3.63	lsp, m
	BC					100	28.81	9.88	2.91	
886-H	AC	30	21.33	10.07	2.12					sp
	AC	100	15.22	5.66	2.69					lsp
	AB	100	17.25	10.85	1.59					lsp
	BC	100	13.80	6.44	2.14					lsp
886-I	AC	30	20.80	11.00	1.89					lsp
	AC	30	20.30	11.40	1.78					lsp
	AB	30	28.70	17.80	1.61					lsp
	AB	30	18.23	11.00	1.66					lsp
	BC	30	22.53	11.17	2.02					lsp
	BC	30	18.90	13.40	1.41					lsp
896-A	AB	30	21.83	14.37	1.52					sp, dol
	AC	100	17.93	7.73	2.31					
896-B	AB	30	27.57	15.47	1.78	50	37.36	17.98	2.07	lsp, m
	AC	30	39.27	6.73	5.83					lsp, m
	*AC	100	32.58	6.89	4.72	50	54.60	7.18	7.60	m Pl. 36
	AB	100	34.34	17.04	2.01					
	BC	100	22.63	7.15	3.16					
896-C	AB	30	18.13	7.43	2.44					sp
896-D	AB	15	10.33	5.93	1.74					sp
896-E	AC	30	13.23	4.03	3.28					
896-G	AB	30	45.80	10.43	4.39					
896-H	BC	30	34.17	11.97	2.85					
8166-A	AC	30	26.73	5.97	4.48					lsp

Sample	Orient.	n				n				Notes
	AC	30	23.13	6.37	3.63	50	36.06	7.66	4.70	lsp
	AC	30	22.77	7.10	3.12					lsp
	AB	30	12.93	10.83	1.19					lsp
	AB	30	23.90	12.47	1.92					lsp
	AB	30	15.60	11.40	1.37					lsp
8166-B	*AC	30	41.00	8.90	4.61	100	54.56	9.84	5.54	sp, dol, m Pl. 24
	AB	30	41.47	17.27	2.40	60	38.45	20.25	1.90	
8166-C	AC	30	18.07	4.60	3.93					m
	AB	30	19.77	6.33	3.12					
8166-D	BC	30	25.10	11.07	2.27	50	26.46	12.12	2.18	lsp/dol
	AC	30	27.27	6.10	4.47	100	14.91	7.40	2.01	lsp/dol
	*AC	30	19.37	9.67	2.00	40	35.67	6.95	5.13	lsp/dol
8166-E	AB	30	15.13	11.83	1.28	98	24.54	11.46	2.14	sp, m, dol Pl. 22
	BC	30	14.30	11.50	1.24	100	20.43	14.24	1.40	sp, m, dol
8166-F	BC	30	22.40	5.83	3.84					sp, m, dol
8166-G	AB	30	18.97	14.20	1.34					sp, dol
8166-H	AC	30	13.17	5.50	2.39					sp, dol
	BC	19	12.84	5.79	2.22					sp, dol
8176-C ⊥	AC	30	28.10	5.20	5.40					def. too intense
	to AC	30	19.77	5.80	3.41					def. too intense
	AB	30	17.23	9.03	1.91					def. too intense
8176-D₁	AC	30	17.17	12.83	1.34					dol
	AB	30	15.97	11.63	1.37					
8176-E	*AC	30	18.03	14.60	1.23	100	21.68	17.76	1.22	dol Pl. 42
	AB	30	19.26	16.63	1.16	50	20.88	18.08	1.15	
8176-G	*AC	30	18.33	12.47	1.47	100	20.60	13.42	1.53	lsp, m
	AB	30	17.97	13.80	1.30	50	11.48	4.98	2.30	Pl. 37
8176-H	BC	30	14.00	11.87	1.18					
	AC	30	14.07	6.23	2.26	100	20.11	15.33	1.31	m
8176-I	AC	30	10.00	5.83	1.71	105	17.08	14.04	1.21	dol

Appendix table—*Continued*

Sample	Cut	# Meas.	Long axis	Short axis	Ratio	# Meas.	Short axis	Long axis	Ratio	Abbreviations + Plates
8176-J	AC	30	10.57	6.97	1.52					dol
8186-A	AC	30	49.70	8.97	5.54					
8186-C$_1$	AC	30	17.40	9.83	1.77					lsp
	AB	30	16.03	12.10	1.32					lsp
	BC	30	13.77	9.27	1.49					lsp
	AC	100	17.64	7.74	2.27					lsp
	AB	100	16.39	12.80	1.28					lsp
	BC	100	12.64	7.65	1.65					lsp
8186-C$_2$	AC	30	17.20	9.43	1.82					sp
	AC	100	18.29	9.96	1.83					sp
	AB	100	17.43	14.37	1.21					
8186-D	AC	30	16.30	3.60	4.53					m
8186-E	AC	30	13.23	6.77	1.96					lsp
	AC	100	13.08	5.99	2.18					lsp
	BC	30	10.50	7.47	1.41					lsp
8186-G	AC	30	21.60	9.07	2.38					sp
	AB	30	24.43	10.33	2.36					sp
	BC	30	12.20	7.00	1.74					sp
8186-I	AC	30	15.77	9.77	1.61					sp
	AB	30	19.27	9.93	1.92					sp
	BC	30	14.10	8.90	1.58					sp
8186-J	AC	30	14.13	11.70	1.21					sp
8186K$_1$	AC	30	25.57	13.37	1.91					sp
	AC	30	14.50	12.80	1.13					sp
8186-K$_2$	AC	30	23.83	10.97	2.17					sp
	AB	30	22.90	10.53	2.17					sp
	AB	30	14.50	8.10	1.79					sp
8186-L	AC	30	11.80	9.07	1.30					sp

Sample	Face	n				n				Notes
8186–M	AC	30	13.67	10.33	1.32					
	AB	30	13.30	11.20	1.19					
	BC	30	12.13	9.33	1.26					sp
8186–N	AB	30	19.28	14.03	1.37					
	AB	30	18.37	12.40	1.48					sp
	AC					30	19.63	13.37	1.46	sp
	AC					30	18.87	13.10	1.44	sp
	AC					30	20.57	13.50	1.52	sp
	AC					30	15.37	10.47	1.47	sp
	BC					30	12.77	11.03	1.16	sp
8186–O	AC	30	12.80	10.07	1.27					sp
	AC	30	63.37	10.63	5.96					sp
	AC	30	36.70	10.67	3.44					sp
8186–P₁	AC	30	13.67	7.23	1.89					sp
	AC	30	12.90	6.47	1.99					sp
	BC	30	12.00	8.00	1.50					sp
8186–P₂	AC	30	14.27	8.03	1.78					sp
	AB	30	14.03	11.77	1.18					sp
8196–A	AC	30	17.50	8.20	2.13					sp
	AC	30	17.70	9.20	1.92					sp
	AB	30	19.27	14.13	1.36					sp
8196–B	AC					30	19.97	9.47	2.11	sp
	AB					30	17.67	12.67	1.39	
	BC					30	13.53	10.43	1.30	
8196–D	AC	30	15.03	9.00	1.67					
8196–E	AC	30	13.17	11.73	1.12					
	AB	30	12.47	11.27	1.10					
8196–H	AC	30	9.53	6.97	1.37					
8196–L	AC	100	14.99	10.25	1.46					sp
1967										
687–A	AC	100	32.21	7.25	4.44	100	20.58	6.42	3.20	sp
687–B	*AC	100	18.35	7.73	2.37	200	31.08	6.80	4.57	lsp, sp, dol, m Pl. 23
	BC									

Appendix table—Continued

Sample	Cut	# Meas.	Long axis	Short axis	Ratio	# Meas.	Short axis	Long axis	Ratio	Abbreviations + Plates
687-B	BC	100	18.65	6.92	2.69	100	20.64	7.69	2.68	
	AB	100	38.09	18.31	2.08	70	42.17	21.08	2.00	
6157-A	AC					100	27.68	7.82	3.53	m
	AC					100	25.65	7.69	3.33	
	BC					100	28.40	9.75	2.91	
6157-B	*AC					100	23.32	10.42	2.23	m Pl. 16
6157-D	AC	100	12.70	9.88	1.28	77	12.53	10.58	1.18	sp, lsp
	AB	100	12.25	9.19	1.33	100	14.22	9.90	1.43	lsp
6157-E	AC	100	10.50	8.35	1.26	100	48.74	30.19	1.61	
6157-F	AC					100	17.93	13.17	1.36	
6157-H	AC					100	29.49	21.16	1.39	sp
6157-J	AC					100	35.51	26.02	1.36	sp
6157-K	AC					100	39.52	30.75	1.28	sp
7117-A	AC	100	14.00	8.11	1.72	100	14.57	8.67	1.68	sp
	AB	100	14.08	8.99	1.56	100	15.14	10.20	1.48	sp
7177-A	AC					100	22.12	6.95	3.18	sp
7177-B	AC					100	19.73	8.31	2.37	sp
	AB					90	20.45	16.24	1.25	sp
7177-E	AC					33	52.24	15.06	3.46	sp
1968										
5288-B	*AC					100	14.28	12.47	1.14	sp Pls. 56, 57
5288-C	AC					100	14.47	12.46	1.16	sp
5288-D	*AC					100	18.77	15.62	1.20	sp Pl. 46
	AB					100	19.21	16.57	1.15	sp
7118-F	AC					100	15.15	13.05	1.16	dol
7118-I	AC					100	15.69	14.01	1.11	dol
7128-A	AC	100	13.14	11.04	1.19					dol

Sample											
7128–B	AC	100	11.62	10.33	1.12	100	18.14	12.37	1.46	dol	
	AC	100	14.30	11.63	1.23	100	17.83	15.20	1.17	dol	
	BC	50	15.56	14.02	1.11	100	15.38	12.49	1.23	sp	
7128–C	AC	100				100	11.18	8.66	1.29	sp	
7158–A	AB	100	12.81	11.27	1.13					sp	
	BC		10.59	8.91	1.18					sp	
7158–B	AC	100	20.17	12.32	1.63					sp	
	AC	100	19.27	16.29	1.18					sp	
	AB	100	16.19	12.72	1.27					sp	
	AC	50	13.86	10.82	1.28					sp	
7158–C	*AC	50				100	17.67	12.77	1.38	dol	Pl. 20
	AB					100	17.97	14.97	1.20	dol	
7168–A	AC	50	12.74	11.36	1.12					sp	
7168–B₁	AC	50	12.96	11.56	1.12					sp	
B₂	AC	100	13.13	11.74	1.11					sp	
BB	AC									sp	
7168–D₁	AC					100	15.09	13.59	1.11	sp	
	AC					100	20.65	17.68	1.16	lsp	
D₂	AC					98	12.30	10.44	1.17	lsp	
7168–E	AC					98	14.44	12.16	1.18	sp	
	AB					99	15.60	12.78	1.22	sp	
7168–F	AC	100	12.13	10.20	1.19					sp	
	AC	100	11.59	9.97	1.16					sp	
	AC	100	9.40	7.49	1.25					sp	
7168–H	AC	50	11.60	10.08	1.15					sp	
7168–I	AC	100	12.68	11.40	1.11					sp	
7168–J	AC	100				100	32.03	26.83	1.19	sp	
7168–K	AC	100	13.81	11.86	1.16					sp	
	AB	100	12.94	11.79	1.09					sp	
7178–B	AC	100	14.40	12.59	1.14					ch	
	AB	50	14.50	12.96	1.11					ch	

227

Appendix table—*Continued*

Sample	Cut	# Meas.	Long axis	Short axis	Ratio	# Meas.	Short axis	Long axis	Ratio	Abbreviations + Plates
7178-C	AB	50	9.28	7.68	1.20	100	13.97	12.12	1.15	sp
	AB	50	11.58	7.24	1.59					
	AC	50	8.68	6.96	1.24	100	14.31	12.37	1.15	sp
7178-E	AC	100	12.90	11.47	1.12					sp
7178-F	AC	50	11.18	9.12	1.22					sp
7178-G	AC	100	14.17	12.48	1.13					sp
7178-H	AC	50	10.76	9.70	1.10					sp
7188-A	AC					100	19.65	11.75	1.67	sp
	AB					100	14.34	11.69	1.22	sp
	BC					100	17.91	12.29	1.45	sp
7198-A	AC					100	14.94	13.31	1.12	sp
	BC					100	15.26	13.41	1.13	sp
	AB					100	15.23	13.72	1.11	sp
7198-B	AC					100	15.13	13.88	1.09	sp
7198-B	AB					100	17.23	15.93	1.08	sp
7198-C	AC					100	10.76	9.20	1.16	sp
7198-D	AC					100	16.81	15.65	1.07	sp
7198-E	BC					100	14.51	12.99	1.11	sp
7198-F_1	AC					100	13.72	12.20	1.12	sp
F_2	AB					98	12.24	10.87	1.12	sp
878-B_1	BC	100	13.82	7.39	1.87	100	11.86	10.14	1.16	sp
B_3	AC	100	23.30	8.73	2.66					sp
	AC	100	27.26	4.09	6.66					sp
	AB	100	19.19	7.48	2.56					sp
9308-E	AC	100	45.90	11.92	3.85					sp
	BC	100	18.69	9.79	1.90					sp
1016-A	AC	50	13.80	3.20	4.31					sp, m

298	AC	35	1.50	100	28.81	19.65	1.46	sp	
302	AC	35	1.14	100	21.60	19.00	1.13	sp	
368–A	AC	35	4.44	100	22.75	5.06	4.49	sp	Pl. 51
344	*AB						1.60		
344	*AC						3.38		Pls. 54, 55
	BC						2.00		
420	*AC	35		35	6.43	4.70	1.31	sp	Pl. 49

Index